Praise for Learn Enough Tutorials

"Going through *Learn Enough Git* is wonderful. I am actually learning... I've done three other Git tutorials and still felt so lost. Doing it all now makes so much sense. It's like a light bulb."
—Janelle Staar

"I bought the *Learn Enough Command Line to Be Dangerous* last fall, and it's paid off sooooo many times in my new job. During my first week, I had a manager sitting right beside me giving me the 'go here, go there, do this, etc.' Having watched, read, and done the exercises, I was confident in getting around the CLI [command-line interface]—and even had him asking, 'What was that shortcut?' For this, I thank you. Now I need a 'Learn even more CLI to be dangerouser.'"
—Thomas Thackery

"I must say, this Learn Enough series is a masterpiece of education. Thank you for this incredible work!"
—Michael King

"I want to thank you for the amazing job you have done with the tutorials. They are likely the best tutorials I have ever read."
—Pedro Iatzky

LEARN ENOUGH DEVELOPER TOOLS TO BE DANGEROUS

Learn Enough Series from Michael Hartl

Visit **informit.com/learn-enough** for a complete list of available publications.

The **Learn Enough** series teaches you the developer tools, Web technologies, and programming skills needed to launch your own applications, get a job as a programmer, and maybe even start a company of your own. Along the way, you'll learn technical sophistication, which is the ability to solve technical problems yourself. And Learn Enough always focuses on the most important parts of each subject, so you don't have to learn everything to get started—you just have to learn enough to be dangerous. The Learn Enough series includes books and video courses so you get to choose the learning style that works best for you.

LEARN ENOUGH DEVELOPER TOOLS TO BE DANGEROUS

Command Line, Text Editor, and Git Version Control Essentials

Michael Hartl

♦♦ Addison-Wesley

Boston • Columbus • New York • San Francisco • Amsterdam • Cape Town
Dubai • London • Madrid • Milan • Munich • Paris • Montreal • Toronto • Delhi • Mexico City
São Paulo • Sydney • Hong Kong • Seoul • Singapore • Taipei • Tokyo

Cover image: Philipp Tur/Shutterstock

Figures 1.1, 1.2, 1.4, 1.6-1.8, 5.4, 5.6-5.9, 5.11, 10.9, 11.8, 11.10, 11.13, A.1: Screenshot © 1995-2021 The Open Group

Figure 1.3: Screenshot © 2021 The Linux Foundation

Figures 3.3, 7.31, 7.32: Screenshot © Regex101

Figures 5.1, 6.1-6.3, 6.10-6.22, 6.24-6.36, 7.1-7.30, 7.33, 7.34, 7.39, 8.3-8.8, 9.1-9.12, 10.3, 11.3-11.7, 11.15-11.17, 11.22-11.24, 11.26: Screenshot © 2021 GitHub, Inc.

Figures 6.4-6.9, 7.35, A.2-A.7: Screenshot © 2021, Amazon Web Services, Inc.

Figures 7.36-7.38: Screenshot © 2020 Wbond

Figure 8.2: Screenshot © 2021 Apple Inc.

Figures 10.2, 10.6, 11.25: Photo of whale, GUDKOV ANDREY/Shutterstock

Figures 11.12, 11.14, 11.19, 11.21, 11.27: Photo of polar bear, Vaclav Sebek/Shutterstock

Many of the designations used by manufacturers and sellers to distinguish their products are claimed as trademarks. Where those designations appear in this book, and the publisher was aware of a trademark claim, the designations have been printed with initial capital letters or in all capitals.

The author and publisher have taken care in the preparation of this book, but make no expressed or implied warranty of any kind and assume no responsibility for errors or omissions. No liability is assumed for incidental or consequential damages in connection with or arising out of the use of the information or programs contained herein.

For information about buying this title in bulk quantities, or for special sales opportunities (which may include electronic versions; custom cover designs; and content particular to your business, training goals, marketing focus, or branding interests), please contact our corporate sales department at corpsales@pearsoned.com or (800) 382-3419.

For government sales inquiries, please contact governmentsales@pearsoned.com.

For questions about sales outside the U.S., please contact intlcs@pearson.com.

Visit us on the Web: informit.com/aw

Library of Congress Control Number: 2022930143

ISBN-13: 978-0-13-784345-9
ISBN-10: 0-13-784345-3

1 2022

Contents

Preface

Learn Enough Developer Tools to Be Dangerous is designed to teach you three essential tools for modern software development: the Unix command line, a text editor, and version control with Git. All three are ubiquitous in the contemporary technology landscape, and yet there are surprisingly few resources for learning them from scratch and seeing how they all fit together. *Learn Enough Developer Tools to Be Dangerous*, which assumes no prerequisites other than general computer knowledge, was created to fill this gap.

The skills you'll learn in this book are valuable whether your interest is in collaborating with developers or becoming a developer yourself. No matter what you want to do—level up in your current job, start a new career, or even start your own company—*Learn Enough Developer Tools to Be Dangerous* is a great place to start.

The individual subjects covered by this book are potentially enormous; entire books can (and have been) written about each of them. But such giant tomes can be overwhelming, especially for beginners, and they generally involve covering many things you don't actually need right away. Instead, this book focuses on the most important aspects of the respective technologies, grounded in the philosophy that you don't have to learn everything to get started—you just have to learn enough to be *dangerous*.

In addition to teaching you specific skills, *Learn Enough Developer Tools to Be Dangerous* also helps you develop *technical sophistication*—the seemingly magical ability to solve practically any technical problem. Technical sophistication includes concrete skills like command lines, text editors, and version control, as well as fuzzier skills like Googling the error message and knowing when to just reboot the darn thing. Throughout this book, there are abundant opportunities to develop technical sophistication in the context of real-world examples.

Finally, although the individual parts of the book are as self-contained as possible, they are also extensively cross-referenced to show how the different tools fit together. You'll learn how to use the command line to launch a text editor, make your changes in the editor, and then return to the command line to record the changes with Git. The result is an integrated introduction to the foundations of software development that's practically impossible to find anywhere else.

Command Line

Part I of *Learn Enough Developer Tools to Be Dangerous*, also known as *Learn Enough Command Line to Be Dangerous*, is an introduction to the Unix command line for complete beginners. It doesn't even assume you know what a "command line" is (though you'll still probably learn a thing or two even if you do). In particular, unlike most command-line tutorials, it doesn't assume you know how to use a text editor (which is the subject of Part II). All of this means you need only basic computer skills (like being able to install new software on your system) to get started.

Like all Learn Enough tutorials, *Learn Enough Command Line to Be Dangerous* is structured as a technical narrative, with each step carefully motivated by real-world uses. Chapter 1 covers the basic notion of a Unix command and shows you how to use your system to learn more about itself. Chapter 2 shows how to use the command line to do things like move, rename, and delete files. Chapter 3 shows how to look inside files (even really big ones), and even how to search through them. Finally, Chapter 4 teaches you how to use the command line to create and navigate directories (folders) to organize files on your system.

The result of finishing *Learn Enough Command Line to Be Dangerous* is a mastery of the basics of a tool that is rarely covered explicitly and yet is everywhere in modern computing. This is especially true of computing in the *Unix tradition*, which includes operating systems like Linux, Android, macOS, and iOS (basically everything but Microsoft Windows, though nowadays even Windows lets you run Linux). This means you'll have a big head start if you're interested in things like web or mobile app development.

Text Editor

Part II, also known as *Learn Enough Text Editor to Be Dangerous*, covers a category of application—known as a *text editor*—that many people don't even know exists, and yet is absolutely essential for professional-grade software development. Text editors are used to make files containing *plain text*, which is the document format used for virtually all Web technologies (like HTML and CSS) and programming languages (JavaScript, Ruby, Python, etc.). As such, knowledge of a text editor is a necessary prerequisite for learning those other important subjects.

Because there is such a wide variety of text editors and user preferences, *Learn Enough Text Editor to Be Dangerous* focuses on the main features shared by virtually all editors. Chapter 5 starts by introducing the powerful Vim text editor, which is available on practically every Unix system in the known universe. Chapter 6 then introduces so-called "modern" text editors, mainly using the free and open-source Atom editor but focusing on features shared with other editors like Sublime Text and Visual Studio Code. As a bonus, this chapter includes an integrated introduction to the popular Markdown formatting language. Chapter 7 then covers more advanced subjects like tab triggers and editing source code, and also shows how to write a *shell script* to extend the capabilities of the command line covered in Part I.

Git

Part III, also known as *Learn Enough Git to Be Dangerous*, covers version control with Git. In line with the approach of the other two parts, *Learn Enough Git to Be Dangerous* doesn't even assume you know what "version control" is (though any familiarity with the subject will still be helpful). As a software system designed to let you track changes in projects, version control might have been considered optional as recently as the early 2000s, but for modern software development it is absolutely essential, and Git has emerged as the clear winner.

Learn Enough Git to Be Dangerous shows how to use Git by tracking changes in a real-world project consisting of a small website (thereby giving you get a head start on web development as well). Chapter 8 shows how to set up a new Git *repository* as a container for your project, beginning with a file consisting of some simple HTML (the markup language of the World Wide Web). Chapter 9 explains how to create a remote backup for your project at GitHub, a popular site for sharing code. Chapter 10 then shows how to use Git to make and record changes to your project, including important techniques known as *branching* and *merging*. Finally, Chapter 11 shows how to use Git to collaborate with other users, including learning how to resolve the kinds of inevitable file conflicts that arise. As a special bonus, you'll learn how to use a free service called GitHub Pages to deploy your site to the live Web.

Additional Features

In addition to the main tutorial material, *Learn Enough Developer Tools to Be Dangerous* includes a large number of exercises to help you test your understanding and to extend the material in the main text. The exercises include frequent hints and often include the expected answers, with community solutions available by separate subscription at www.learnenough.com.

For completeness, *Learn Enough Developer Tools to Be Dangerous* includes an appendix on setting up a development environment, including instructions for native systems (macOS, Linux, Windows) and a preconfigured cloud IDE (integrated development environment). This material is also available for free online at www.learnenough.com/dev-environment, which can be consulted for the most up-to-date instructions.

Final Thoughts

Learn Enough Developer Tools to Be Dangerous is designed as a foundational text for modern software development. After learning the developer tools covered in this tutorial, and especially after beginning to develop your technical sophistication, you'll be ready for a huge variety of other resources, including books, blog posts, and online documentation. You'll also have the prerequisites needed for the other Learn Enough tutorials: *Learn Enough HTML, CSS and Layout to Be Dangerous*, *Learn Enough JavaScript to Be Dangerous*, and *Learn Enough Ruby to Be Dangerous*. You can even go on to learn professional-grade web development with the *Ruby on Rails™ Tutorial*.

Learn Enough Scholarships

Learn Enough is committed to making a technical education available to as wide a variety of people as possible. As part of this commitment, in 2016 we created the Learn Enough Scholarship program (https://www.learnenough.com/scholarship). Scholarship recipients get free or deeply discounted access to the Learn Enough All Access subscription, which includes all of the Learn Enough online book content, embedded videos, exercises, and community exercise answers.

As noted in a 2019 RailsConf Lightning Talk (https://youtu.be/AI5wmnzzBqc?t =1076), the Learn Enough Scholarship application process is incredibly simple: just fill out a confidential text area telling us a little about your situation. The scholarship criteria are generous and flexible—we understand that there are an enormous number of reasons for wanting a scholarship, from being a student, to being between jobs, to living in a country with an unfavorable exchange rate against the U.S. dollar. Chances are that, if you feel like you've got a good reason, we'll think so, too.

So far, Learn Enough has awarded more than 2,500 scholarships to aspiring developers around the country and around the world. To apply, visit the Learn Enough Scholarship page at www.learnenough.com/scholarship. Maybe the next scholarship recipient could be you!

Register your copy of *Learn Enough Developer Tools to Be Dangerous* on the InformIT site for convenient access to updates and/or corrections as they become available. To start the registration process, go to informit.com/register and log in or create an account. Enter the product ISBN (9780137843459) and click Submit. Look on the Registered Products tab for an Access Bonus Content link next to this product, and follow that link to access any available bonus materials. If you would like to be notified of exclusive offers on new editions and updates, please check the box to receive email from us.

About the Author

Michael Hartl (https://www.michaelhartl.com/) is the creator of the *Ruby on Rails Tutorial* (https://www.railstutorial.org/), one of the leading introductions to web development, and is cofounder and principal author at Learn Enough (https://www.learnenough.com/). Previously, he was a physics instructor at the California Institute of Technology (Caltech), where he received a Lifetime Achievement Award for Excellence in Teaching. He is a graduate of Harvard College, has a Ph.D. in Physics from Caltech, and is an alumnus of the Y Combinator entrepreneur program.

PART I

Command Line

CHAPTER 1

Basics

Welcome to *Learn Enough Developer Tools to Be Dangerous*! This tutorial is designed to teach you three essential tools—the *command line*, a *text editor*, and *version control*—needed for modern software development (or, as I prefer to call it, "computer magic" (Box 1.1)). It is aimed both at those who work with software developers and those who aspire to become developers themselves.

Box 1.1: The Magic of Computers

Computers may be as close as we get to *magic* in the real world: We type incantations into a machine, and—if the incantations are right—the machine does our bidding. To perform such magic, computer witches and wizards rely not only on words, but also on wands, potions, and an ancient tome or two. Taken together, these tricks of the trade are known as *software development*: computer programming, plus tools like *command lines*, *text editors*, and *version control*. Knowledge of these tools is perhaps the main dividing line between "technical" and "non-technical" people (or, to put it in magical terms, between magicians and Muggles). The present tutorial represents the first steps needed to cross this technical/non-technical divide. The resulting *technical sophistication* (Box 1.4) will make us software magicians—able to cast computer spells, and get the machine to do our bidding.

Learn Enough Developer Tools to Be Dangerous consists of three parts, each covering one of the essential tools (and based on the corresponding online course from Learn Enough https://www.learnenough.com/):

- Part I: *Learn Enough Command Line to Be Dangerous* (https://www.learnenough .com/command-line) on the Unix command line

- Part II: *Learn Enough Text Editor to Be Dangerous* (https://www.learnenough .com/text-editor) on how to use a text editor

- Part III: *Learn Enough Git to Be Dangerous* (https://www.learnenough.com/git) on version control with Git

Because of the close association between the parts and the standalone courses, *Learn Enough Developer Tools to Be Dangerous* will use terms like "Part I" and "*Learn Enough Command Line to Be Dangerous*" interchangeably.

Part I of *Learn Enough Developer Tools to Be Dangerous* is an introduction to the Unix command line for complete beginners, so it doesn't assume any background other than general computer knowledge (how to launch an application, how to use a web browser, how to touch type, etc.). Each subsequent part of *Learn Enough Developer Tools to Be Dangerous* assumes only the material in the preceding parts, so to get started you don't even need to know what "command line", "text editor", and "version control" mean. Moreover, even if you are already somewhat familiar with one or more of these subjects, following this tutorial (and doing the exercises) will help fill in any gaps in your knowledge, and might even teach you a few new things.

Many programming tutorials either gloss over developer tools or assume you already know how to use them. But understanding the basics of the command line, text editors, and version control is *absolutely essential* to becoming a skilled developer.[1] Indeed, if you look at the desktop of an experienced computer programmer, even on a system with a polished graphical user interface like macOS, you are likely to find a large number of "terminal windows", text-editor windows, and version-control commands (Figure 1.1). Proficiency with the developer tools covered in *Learn Enough Developer Tools to Be Dangerous* is also useful for anyone who needs to *work* with developers, such as product managers, project managers, and designers. Making this essential component of technical sophistication accessible to as broad an audience as possible is the goal of this tutorial.

---

1. Regarding the version of the command line covered in this tutorial, this statement applies specifically to the *Unix tradition*, which is the principal computing tradition behind the Internet and the World Wide Web. In addition to being important for developers, knowing the command line is also essential for system administrators (sysadmins).

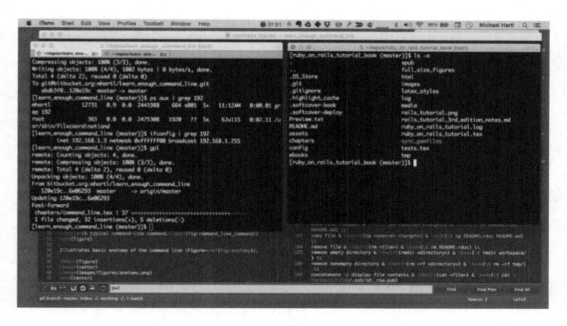

Figure 1.1: Terminal windows on the desktop of an experienced developer.

1.1   Introduction

As author Neal Stephenson famously put it, "In the Beginning… Was the Command Line" (paperback edition, Avon Books, 1999). Although a graphical user interface (GUI) can dramatically simplify computer use, in many contexts the most powerful and flexible way to interact with a computer is through a *command-line interface* (CLI). In such an interface, the user types *commands* that tell the computer to perform desired tasks. These commands can then be combined in various ways to achieve a variety of outcomes. An example of a typical command–line command appears in Figure 1.2.

This tutorial covers the basics of the Unix command line, where *Unix* refers to a family of operating systems that includes Linux, Android, iOS (iPhone and iPad),

```
[projects]$ rm -f foo.txt
```

Figure 1.2: A prototypical command-line command.

and macOS.[2] Unix systems serve most of the software on the World Wide Web, run most mobile and tablet devices, and power many of the world's desktop computers as well. As a result of Unix's central role in modern computing, this tutorial covers the Unix way of developing software. The main exception to Unix's dominance is Microsoft Windows, which is not part of the Unix tradition, but those who mostly develop using native Windows development tools will still benefit from learning the Unix command line. Among other things, at some point such users are likely to need to issue commands on a Unix server (e.g., via the "secure shell" command **ssh**), at which point familiarity with Unix commands becomes essential. As a result, Microsoft Windows users are encouraged to set up a Linux-compatible development environment by following the steps in Section A.3.3 or by using the Linux-based cloud IDE discussed in Section A.2.

1.2 Running a Terminal

To run a command-line command, we first need to start a *terminal*, which is the program that gives us a command line. The exact details depend on the particular operating system you're using.

macOS
On macOS, you can open a terminal window using the macOS application *Spotlight*, which you can launch either by typing ⌘␣ (Command-space) or by clicking on the magnifying glass in the upper right part of your screen. Once you've launched Spotlight, you can start a terminal program by typing "terminal" in the Spotlight Search bar. (If you are interested in using a more advanced and customizable terminal program, I recommend installing iTerm, but this step is optional.)

At this point, you might see the alert shown in Listing 1.1.

Listing 1.1: A macOS terminal alert.

```
The default interactive shell is now zsh.
To update your account to use zsh, please run `chsh -s /bin/zsh`.
For more details, please visit https://support.apple.com/kb/HT208050.

[~]$
```

2. In a fairly typical turn of events, the name *Unix* started as a pun on a rival system called *Multics*.

This alert is the result of a change made in macOS Catalina. You don't need to do anything about it right now; we'll address this issue the first time it makes any difference in this tutorial (Section 2.3). For more information, see the Learn Enough blog post "Using Z Shell on Macs with the Learn Enough Tutorials" (https://news.learnenough.com/macos-bash-zshell).

Linux

On Linux, you can click the terminal icon as shown in Figure 1.3. The result should be something like Figure 1.4, although the exact details on your system will likely differ.

Windows

On Windows, the recommended option is to install Linux (which, incredibly, Microsoft has decided to support natively) as described in the steps in Section A.3.3.

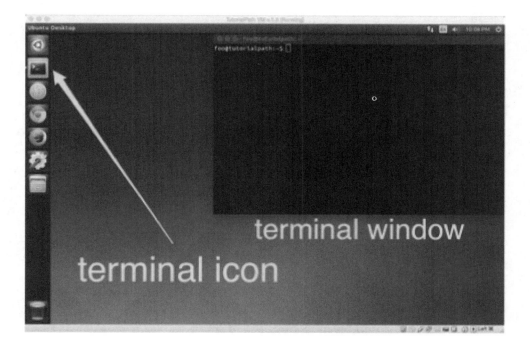

Figure 1.3: The Linux terminal icon.

Figure 1.4: A terminal window.

Once Linux is installed, you should look for a terminal icon as described in Section 1.2. Apply your technical sophistication (Box 1.4) if you get stuck.

Terminal Window

Regardless of which operating system you use, your terminal window should look something like Figure 1.4, though details may differ.

The example we saw in Figure 1.2 includes all of the typical elements of a command, as illustrated in Figure 1.5: the *prompt* (to "prompt" the user to do something) followed by a *command* (as in "give the computer a command"), an *option* (as in "choose a different option"),[3] and an *argument* (as in the "argument of a function" in mathematics). It's essential to understand that the prompt is supplied automatically by the terminal, and you do not need to type it. (Indeed, if you do type it, it will likely

---

3. An option is sometimes also called a *flag*.

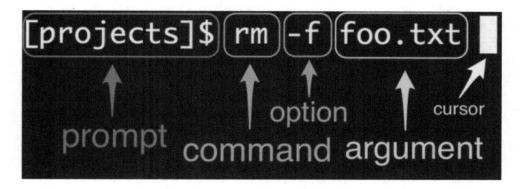

Figure 1.5: Anatomy of a command line. (Your prompt may differ.)

result in an error.) Moreover, the exact details of the prompt will differ, and are not important for the purposes of this tutorial (Box 1.2).

Box 1.2: What Is the Prompt?

Every command line starts with some symbol or symbols designed to "prompt" you to action. The prompt usually ends with a dollar sign $ or a percent sign %, and is preceded by information that depends on the details of your system. For example, on some systems the prompt might look like this:

```
Michael's MacBook Air:~ mhartl$
```

In Figure 1.4, the prompt looks like this instead:

```
[~]$
```

and in Figure 1.5 it looks like this:

```
[projects]$
```

Finally, the prompt I'm looking at right now looks like this:

```
[learn_enough_command_line (first-draft)]$
```

For the purposes of this tutorial, the details of the prompt are not important, but we will discuss useful ways to customize the prompt in Part II.

1.2.1 Exercises

Learn Enough Developer Tools to Be Dangerous includes a large number of exercises. I strongly recommend getting in the habit of attempting them before moving on to the next section, as they reinforce the material we've just covered and will give you essential practice in using the many commands discussed. It's not generally the case that they are *required* to proceed, though, so if you get stuck it's sometimes a good idea to continue forward and then revisit the exercise at a later time. Indeed, this is good advice for the main text as well—you'll be surprised how often a seemingly impossible idea or intractable problem will look easy the second time around.

1. By referring to Figure 1.5, identify the prompt, command, options, arguments, and cursor in each line of Figure 1.6.

2. Most modern terminal programs have the ability to create multiple *tabs* (Figure 1.7), which are useful for organizing a set of related terminal windows.[4] By examining the menu items for your terminal program (Figure 1.8), figure out how to create a new tab. *Extra credit*: Learn the keyboard shortcut for creating a new tab. (Learning keyboard shortcuts for your system is an excellent habit to cultivate.)

1.3 Our First Command

We are now prepared to run our first command, which prints the word "hello" to the screen. (The place where characters get printed is known as "standard out", which is usually just the screen, and rarely refers to a physical printer.) The command is **echo**, and the argument is the string of characters—or simply *string* for short—that we want to print. To run the **echo** command, type "echo hello" at the prompt, and then press the Return key (also called Enter):

```
$ echo hello
hello
$
```

---

4. For example, when developing web applications, in addition to my main command-line tab I often have separate tabs for running a local web server (https://www.railstutorial.org/book/beginning#sec-rails_server) and an automated test suite (https://www.railstutorial.org/book/static_pages#sec-our_first_test).

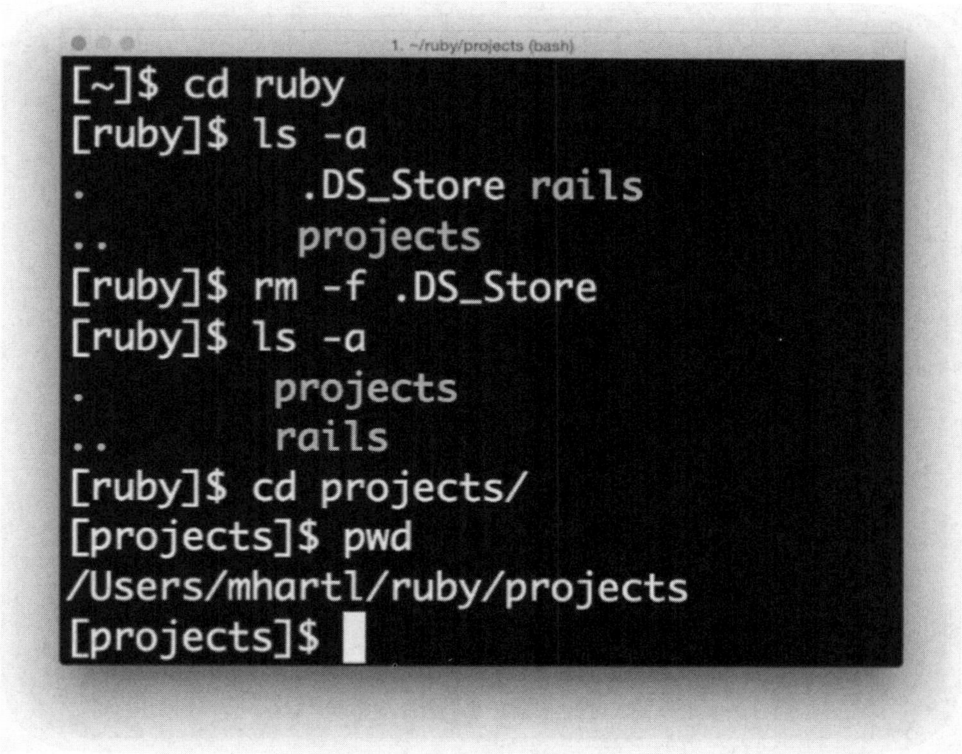

Figure 1.6: A series of typical commands.

(I recommend always typing the commands out yourself, which will let you learn more than if you rely on copying and pasting.) Here we see that **echo hello** prints "hello" and then returns another prompt. Note that, for brevity, I've omitted all characters in the prompt except the dollar sign **$**.

Just to make the pattern clear, let's try a second **echo** command:

```
$ echo "goodbye"
goodbye
$ echo 'goodbye'
goodbye
$
```

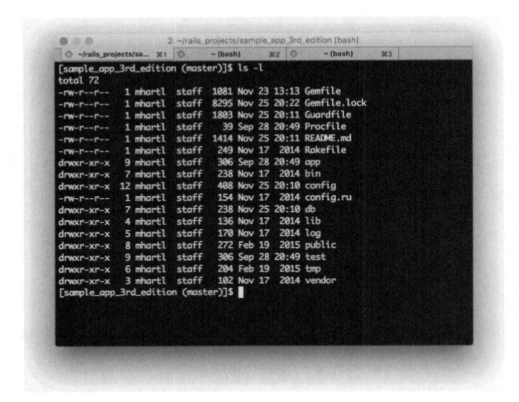

Figure 1.7: A terminal window with three tabs.

Note here that we've wrapped "goodbye" in quotation marks—and we also see that we can use either double quotes, as in **"goodbye"**, or single quotes, as in **'goodbye'**. Such quotes can be used to group strings visually, though in many contexts they are not required by **echo** (Listing 1.2).[5]

Listing 1.2: Printing "hello, goodbye" two different ways.

```
$ echo hello, goodbye
hello, goodbye
$ echo "hello, goodbye"
hello, goodbye
$
```

---

5. There are subtle differences between the two cases, but they aren't important at the level of this tutorial. Use your technical sophistication (Box 1.4) if you're curious. *Hint*: Google searches are your friends.

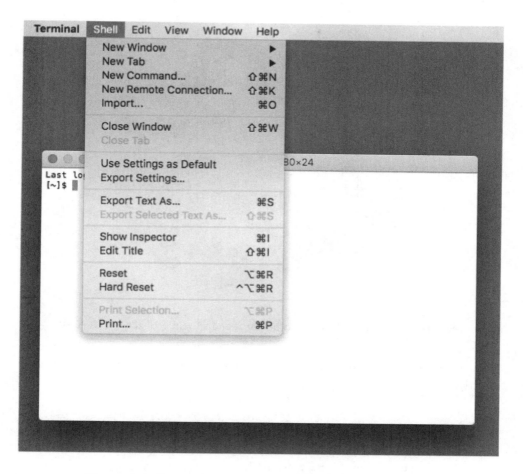

Figure 1.8: Some menu items for the default macOS terminal.

One thing that can happen when using quotes is accidentally not matching them, as follows:

```
$ echo "hello, goodbye
>
```

At this point, it seems we're stuck. There are specific ways out of this quandary (in fact, in this case you can just add a closing quote and hit return), but it's good to

Figure 1.9: This cat appears to be stuck and should probably hit `Ctrl-C`.

have a general strategy for getting out of trouble (Figure 1.9).[6] This strategy is called "Ctrl–C" (Box 1.3).

Box 1.3: Getting Out of Trouble

When using the command line, there are lots of things that can get you in trouble, by which I mean the terminal will just hang or otherwise end up in a state that makes entering further commands difficult or impossible. Here are some examples of such commands:

```
$ echo "hello
```

```
$ grep foobar
```

```
$ yes
```

---

6. Image courtesy of Akitameldes/Shutterstock.

```
$ tail

$ cat
```

In every case, the solution is the same: hit `Ctrl-C` (pronounced "control-see"). Here `Ctrl` refers to the "control" key on your keyboard, and `C` refers to the key labeled "C". `Ctrl-C` thus means "While holding down the control key, press C." In particular, `C` does *not* refer to the capital letter C, so you should not press Shift in addition to Ctrl. (`Ctrl-C` sends a *control code* to the terminal and has nothing to do with producing a capital C when typing normal text.) The result of typing `Ctrl-C` is sometimes written as ^C, like this:

```
$ tail
^C
```

The origins of `Ctrl-C` are somewhat obscure, but as a mnemonic I like to think of it as meaning "cancel". However you remember it, *do* remember it: When you get into trouble at the command line, your best bet is usually to hit `Ctrl-C`. *Note*: When `Ctrl-C` fails, 90% of the time hitting ESC (escape) will do the trick.

1.3.1   Exercises

1. Write a command that prints out the string "hello, world". *Extra credit*: As in Listing 1.2, do it two different ways, both with and without using quotation marks.

2. Type the command **echo 'hello** (with a mismatched single quote), and then get out of trouble using the technique from Box 1.3.

1.4   Man Pages

The program we're using to run a command line, which is technically known as a *shell*,[7] includes a powerful (though often cryptic) tool to learn more about available commands. This tool is itself a command-line command called **man** (short for

---

7. Many introductions to the command line cover elements of the shell that require knowledge of a text editor—knowledge which (as noted in the introduction) this tutorial does *not* assume. As a result, we'll defer these important topics to the follow-on tutorials to this one (Section 4.6), starting with *Learn Enough Text Editor to Be Dangerous*.

"manual"), and we use it by typing **man** and then the name of the command we want to learn more about:

```
$ man echo
```

The result is to print out a so-called *man page* for the command in question (in this case, **echo**). The details are system–dependent, but on my system the result of running **man echo** appears as in Listing 1.3.

Listing 1.3: The result of running **man echo**.

```
$ man echo
ECHO(1)               BSD General Commands Manual          ECHO(1)

NAME
   echo -- write arguments to the standard output

SYNOPSIS
   echo [-n] [string ...]

DESCRIPTION
   The echo utility writes any specified operands, separated by single blank
   (` ') characters and followed by a newline (`\n') character, to the stan-
   dard output.

   The following option is available:

   -n  Do not print the trailing newline character. This may also be
       achieved by appending `\c' to the end of the string, as is done by
       iBCS2 compatible systems. Note that this option as well as the
       effect of `\c' are implementation-defined in IEEE Std 1003.1-2001
       (``POSIX.1'') as amended by Cor. 1-2002. Applications aiming for
       maximum portability are strongly encouraged to use printf(1) to
       suppress the newline character.
:
```

On the last line of Listing 1.3, note the presence of a colon **:**, which indicates that there is more information below. The details of this last line are also system-dependent, but on any system you should be able to access subsequent information one line at a time by pressing the down arrow key, or one page at a time by pressing the spacebar. To exit the man page, press "q" (for "quit"). (This interface to the man pages is the same as for the **less** program, which we'll learn about in Section 3.3.)

Figure 1.10: Applying **man** to **man**.

Because **man** itself is a command, we can apply **man** to **man** (Figure 1.10),[8] as shown in Listing 1.4.

Listing 1.4: The result of running **man man**.

```
$ man man
man(1)                                          man(1)

NAME
    man - format and display the on-line manual pages

SYNOPSIS
    man [-acdfFhkKtwW] [--path] [-m system] [-p string] [-C config_file]
    [-M pathlist] [-P pager] [-B browser] [-H htmlpager] [-S section_list]
    [section] name ...
```

8. Image courtesy of Monkey Business Images/Shutterstock.

```
DESCRIPTION
    man formats and displays the on-line manual pages. If you specify sec-
    tion, man only looks in that section of the manual. name is normally
    the name of the manual page, which is typically the name of a command,
    function, or file. However, if name contains a slash (/) then man
    interprets it as a file specification, so that you can do man ./foo.5
    or even man /cd/foo/bar.1.gz.

    See below for a description of where man looks for the manual page
    files.

OPTIONS
    -C config_file
  :
```

We can see from Listing 1.4 that the synopsis of **man** looks something like this:

```
man [-acdfFhkKtwW] [--path] [-m system] [-p string] ...
```

This is what I meant above when I described man pages as "often cryptic". Indeed, in many cases I find the details of man pages to be almost impossible to understand, but being able to scan over the man page to get a high-level overview of a command is a valuable skill, one well worth acquiring. To get used to reading man pages, I recommend running **man <command name>** when encountering a new command. Even if the details aren't entirely clear, reading the man pages will help develop the valuable skill of *technical sophistication* (Box 1.4).

Box 1.4: Technical Sophistication

In mathematics, many subjects can be developed by applying pure deduction to a small number of assumptions, or *axioms*; examples include algebra, geometry, number theory, and analysis. As a result, such subjects are completely self-contained, and thus have no formal prerequisites—in principle, even a small child could learn them. In practice, though, something else is required, and mathematicians often recommend the informal prerequisite of *mathematical maturity*, which consists of the experience and general sophistication needed to understand and write mathematical proofs.

In technology, a similar skill (or, more accurately, set of skills) exists in the form of *technical sophistication*. In addition to "hard skills" like familiarity with text editors and the Unix command line, technical sophistication includes "soft skills" like looking for promising menu items and knowing the kinds of search terms to drop into Google (as illustrated in "Tech Support Cheat Sheet" (https://m.xkcd.com/627/) from xkcd), along with an *attitude* of doing what it takes to make the machine do our bidding (Box 1.1).

These soft skills, and this attitude, are hard to teach directly, so as you progress through this and subsequent Learn Enough tutorials you should always be on the lookout for opportunities to increase your technical sophistication (such as, for example, learning how to get the gist of a program by scanning its man page (Section 1.4)). Over time, the cumulative effect will be that, like the author of "Tech Support Cheat Sheet", you'll have the seemingly magical ability to do everything in every program.

By the way, "Tech Support Cheat Sheet" is missing a few important techniques for solving common problems (listed here in increasing order of severity, which is the order in which you should try them):

1. "Have you restarted the application?"

2. "Have you rebooted the device?" or (closely related) "Have you turned it off, waited 30 seconds, and turned it on again?" (This is known as *power cycling*.)

3. "Have you tried uninstalling and reinstalling the app?"

Step #2 alone probably solves 90% of unexplained computer problems.

1.4.1 Exercises

1. According to the man page, what are the official short and long descriptions of **echo** on your system?

2. As seen in Listing 1.2, by default the **echo** command prints its argument to the screen and then puts the new prompt on a new line. The way it does this is by appending a special character called a *newline* that literally puts the string on a new line. (The newline character is usually written as **\n**, pronounced "backslash n".) Because **echo** is often used in programs to print out a sequence of strings *not* separated by newlines, there is a special command–line option to prevent the newline from being inserted.

   By reading the man page for **echo**, determine the command needed to print out "hello" *without* the trailing newline, and verify using your terminal that it works as expected. *Hints*: To determine the placement of the command-line option, it may help to refer to Figure 1.5. By comparing your result with Listing 1.5 and Listing 1.6, you should be able to verify that you've used the option properly. (*Note*: This exercise may fail when using the default terminal program on some older versions of macOS. In this case, I recommend installing iTerm (which isn't a bad idea anyway).)

Listing 1.5: The result of running **echo** with a newline (without option).

```
hello
[~]$
```

Listing 1.6: The result of running **echo** without a newline (with option).

```
hello[~]$
```

1.5 Editing the Line

Command lines include several features to make it easy to repeat previous commands, possibly in edited form. These and many other command-line features often involve special keys on the keyboard, so for reference Table 1.1 shows these symbols for the various keys on a typical Macintosh keyboard. Apply your technical sophistication (Box 1.4) if your keyboard differs.

Table 1.1: Miscellaneous keyboard symbols.

Key	Symbol
Command	⌘
Control	^
Shift	⇧
Option	⌥
Up, down, left, right	↑ ↓ ← →
Enter/Return	↵
Tab	⇥
Delete	⌫

One of the most useful ways to edit the line is to "up arrow" ↑, which simply retrieves the previous command. Pressing up arrow again moves further up the list of commands, while "down arrow" ↓ goes back toward the bottom.

Other common ways to edit the line use the control key, which (as we saw in Box 1.3) is usually written as **Ctrl** or **^**. For example, when typing a new command, or dealing with a previous command, it is often convenient to be able to move quickly within the line. Suppose we typed

```
$ goodbye
```

only to realize that we wanted to put **echo** in front of it. We could use the left arrow key ← to get to the beginning of the line, but it's easier to type **^A**, which takes us there immediately. Similarly, **^E** moves to the end of the line.[9] Finally, **^U** clears to the beginning of the line and lets us start over.

The combination of **^A**, **^E**, and **^U** will work on most systems, but they don't do you much good if you're editing a longer line, such as this one containing the first line of Sonnet 1 by William Shakespeare (Listing 1.7).

Listing 1.7: Printing the first line of Shakespeare's first sonnet.

```
$ echo "From fairest creatures we desire increase,"
```

Suppose we wanted to change "From" to "FRom" to more closely match the text from the original sonnet (Figure 1.11).[10] We could type **^A** followed by the right arrow key a few times, but on some systems it's possible to move directly to the desired spot by combining the keyboard and mouse via Option-click. That is, you can hold down the Option key on your keyboard (if it exists),[11] and then click with the mouse pointer

---

9. There are also commands for moving one *word* at a time (the keyboard sequences **ESC F** and **ESC B**), but I hardly ever use them myself, so it's clear they are not required to be *dangerous*.

10. Note that the Original Pronunciation (OP) of Shakespearean English is different from modern pronunciation. Generally speaking, Shakespeare's sonnets include many word pairs that don't rhyme in modern English but do in OP. In the case of Figure 1.11, the word "memory" should be pronounced "MEM-or-aye", leading to a rhyme between lines 2 and 4 (ending in *die* and *memory*, respectively).

11. Some keyboards lack an Option key, so obviously this trick won't work on such systems.

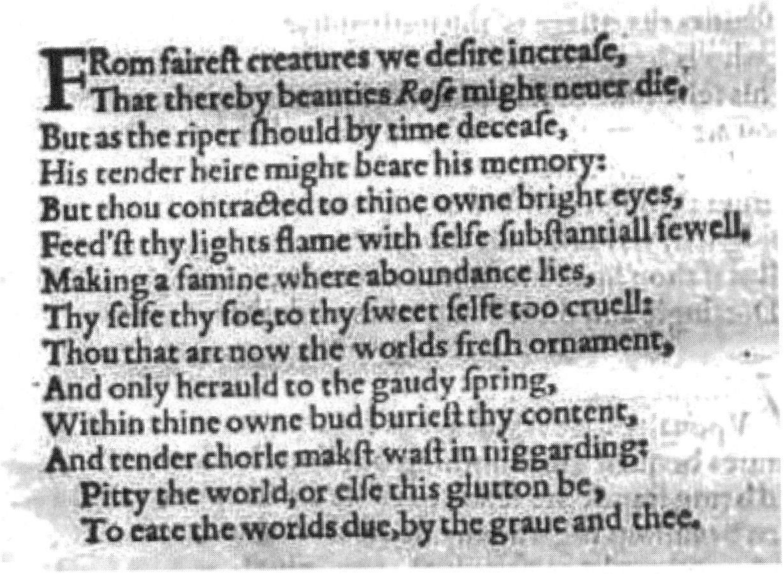

FRom faireſt creatures we deſire increaſe,
That thereby beauties *Roſe* might neuer die,
But as the riper ſhould by time deceaſe,
His tender heire might beare his memory:
But thou contraꞔted to thine owne bright eyes,
Feed'ſt thy lights flame with ſelfe ſubſtantiall fewell,
Making a famine where aboundance lies,
Thy ſelfe thy foe,to thy ſweet ſelfe too cruell:
Thou that art now the worlds freſh ornament,
And only herauld to the gaudy ſpring,
Within thine owne bud burieſt thy content,
And tender chorle makſt waſt in niggarding:
    Pitty the world,or elſe this glutton be,
    To eate the worlds due,by the graue and thee.

Figure 1.11: The original appearance of Shakespeare's first sonnet.

on the place in the command where you want the cursor. This would let us move right to the "o" in "From", allowing us to delete the "r" and yielding Listing 1.8 directly.

Listing 1.8: The result of editing a longer command-line command.

```
$ echo "FRom fairest creatures we desire increase,"
```

I usually move around the command line with a combination of ^A, ^E, and right and left arrow keys, but for longer commands Option-click can be a big help. (I also frequently change my mind about the exact command I'm typing, in which case I usually find that hitting ^U and starting over again is the fastest way to proceed.)

1.5.1 Exercises

1. Using the up arrow, print to the screen the strings "fee", "fie", "foe", and "fum" without retyping **echo** each time.

2. Starting with the line in Listing 1.7, use any combination of **^A**, **^E**, arrow keys, or Option–click to change the occurrences of the short s to the archaic long s "ſ" in order to match the appearance of the original (Figure 1.11). In other words, the argument to **echo** should read "FRom fairest creatures we desire increase,". *Hint*: It's unlikely that your keyboard can produce "s" natively, so either copy it from the text of this tutorial or Google for it and copy it from the Internet. (If you have trouble copying and pasting into your terminal, I suggest applying the ideas in Box 1.4 to figure out how to do it on your system.)

1.6   Cleaning Up

When using the command line, sometimes it's convenient to be able to clean up by clearing the screen, which we can do with **clear**:

```
$ clear
```

A keyboard shortcut for this is **^L**.

Similarly, when we are done with a terminal window (or tab) and are ready to exit, we can use the **exit** command:

```
$ exit
```

A keyboard shortcut for this is **^D**.

1.6.1 Exercises

1. Clear the contents of the current tab.

2. Open a new tab, execute **echo 'hello'**, and then exit.

1.7 Summary

Important commands from this chapter are summarized in Table 1.2.

Table 1.2: Important commands from Chapter 1.

Command	Description	Example
echo \<string\>	Print string to screen	$ echo hello
man \<command\>	Display manual page for command	$ man echo
^C	Get out of trouble	$ tail ^C
^A	Move to beginning of line	
^E	Move to end of line	
^U	Delete to beginning of line	
Option-click	Move cursor to location clicked	
Up & down arrow	Scroll through previous commands	
clear or ^L	Clear screen	$ clear
exit or ^D	Exit terminal	$ exit

1.7.1 Exercises

1. Write a command to print the string **Use "man echo"**, *including* the quotes; i.e., take care not to print out **Use man echo** instead. *Hint*: Use double quotes in the inner string, and wrap the whole thing in single quotes.

2. By running **man sleep**, figure out how to make the terminal "sleep" for 5 seconds, and execute the command to do so.

3. Execute the command to sleep for 5000 seconds, realize that's well over an hour, and then use the instructions from Box 1.3 to get out of trouble.

CHAPTER 2
Manipulating Files

Having covered how to run a basic command, we're now ready to learn how to manipulate files, one of the most important tasks at the command line. Because Part I assumes no technical prerequisites, we're not going to require any familiarity with programs designed to edit text. (As noted in Chapter 1, such programs, called *text editors*, are the subject of Part II.) This means that we'll need to create files by hand at the command line. But this is a feature, not a bug (Box 2.1), because learning to create files at the command line is a valuable skill in itself.

Box 2.1: Learning to Speak "Geek"

One important part of learning software development is becoming familiar with the hacker, nerd, and geek culture from which much of it springs. For example, the phrase "It's not a bug, it's a feature" is a common way of recasting a seeming flaw as a virtue. For example, if a user encounters what looks like an odd bit of behavior in a program and brings it to the attention of the developer, the developer might reply, "It's not a bug, it's a feature!"

The Jargon File, which includes an enormous and entertaining lexicon of hacker terms, expands on this theme in its entry on *feature*:

"Undocumented feature" is a common, allegedly humorous euphemism for a *bug*. There's a related joke that is sometimes referred to as the "one-question geek test". You say to someone "I saw a Volkswagen Beetle today with a vanity license plate that read FEATURE". If he/she laughs, he/she is a *geek*.

Figure 2.1: It's not a bug, it's a feature.

The joke here is that, because "bug" is a common slang term for a Volkswagen Beetle, a Beetle with the vanity plate FEATURE is a real-life manifestation of "It's not a bug, it's a feature" (Figure 2.1).[1]

  Even if you're not a geek or nerd yourself, learning to "speak geek" will help you navigate both the technological landscape and the social world that surrounds it.

2.1 Redirecting and Appending

Let's pick up (more or less) where we left off in Chapter 1, with an **echo** command to print out the first line of Shakespeare's first sonnet (Listing 1.7):

---

1. Image courtesy of Wirestock Creators/Shutterstock.

```
$ echo "From fairest creatures we desire increase,"
From fairest creatures we desire increase,
```

Our task now is to create a file containing this line. Even without the benefit of a text editor, it is possible to do this using the *redirect operator* **>**:

```
$ echo "From fairest creatures we desire increase," > sonnet_1.txt
```

(Recall that you can use up arrow to retrieve the previous command rather than typing it from scratch.) Here the right angle bracket **>** takes the string output from **echo** and redirects its contents to a file called **sonnet_1.txt**.

How can we tell if the redirect worked? We'll learn some more advanced command-line tools for inspecting files in Chapter 3, but for now we'll use the **cat** command, which simply dumps the contents of the file to the screen:

```
$ cat sonnet_1.txt
From fairest creatures we desire increase,
```

The name **cat** is short for "concatenate", which is a hint that it can be used to combine the contents of multiple files, but the usage above (to dump the contents of a single file to the screen) is extremely common. Think of **cat** as a "quick-and-dirty" way to view the contents of a particular file (Figure 2.2).[2]

In order to add the second line of the sonnet (in modernized spelling), we can use the *append operator* **>>** as follows:

```
$ echo "That thereby beauty's Rose might never die," >> sonnet_1.txt
```

This just adds the line to the end of the given file. As before, we can see the result using **cat**:

```
$ cat sonnet_1.txt
From fairest creatures we desire increase,
That thereby beauty's Rose might never die,
```

---

2. Image courtesy of garetsworkshop/Shutterstock.

Figure 2.2: Viewing a file with `cat`.

(To get to this command, I hope you just hit up arrow twice instead of retyping it. If so, you're definitely getting the hang of this.) The result above shows that the double right angle bracket **>>** appended the string from **echo** to the file **sonnet_1.txt** as expected.

Modernized treatments of the *Sonnets* sometimes emend *Rose* to *rose* (thereby obscuring the likely meaning), and we can make a second file following this convention using two more calls to **echo**:

```
$ echo "From fairest creatures we desire increase," > sonnet_1_lower_case.txt
$ echo "That thereby beauty's rose might never die," >> sonnet_1_lower_case.txt
```

In order to facilitate the comparison of files that are similar but not identical, Unix systems come with the helpful **diff** command:

```
$ diff sonnet_1.txt sonnet_1_lower_case.txt
< That thereby beauty's Rose might never die,
---
> That thereby beauty's rose might never die,
```

When discussing computer files, *diff* is frequently employed both as a noun ("What's the diff between those files?") and as a verb ("You should diff the files to see what changed."). As with many technical terms, this sometimes bleeds over into common usage, such as "Diff present ideas against those of various past cultures, and see what you get."[3]

2.1.1   Exercises

At the end of each of the exercises below, use the **cat** command to verify your answer.

1. Using **echo** and **>**, make files called **line_1.txt** and **line_2.txt** containing the first and second lines of Sonnet 1, respectively.

2. Replicate the original **sonnet_1.txt** (containing the first two lines of the sonnet) by first redirecting the contents of **line_1.txt** and then appending the contents of **line_2.txt**. Call the new file **sonnet_1_copy.txt**, and confirm using **diff** that it's identical to **sonnet_1.txt**. *Hint*: When there is no diff between two files, **diff** simply outputs nothing.

3. Use **cat** to combine the contents of **line_1.txt** and **line_2.txt** in reverse order using a *single* command, yielding the file **sonnet_1_reversed.txt**. *Hint*: The **cat** command can take multiple arguments.

---

3. From the essay "What You Can't Say" (http://www.paulgraham.com/say.html) by Paul Graham (2004). As Graham notes:

> The verb "diff" is computer jargon, but it's the only word with exactly the sense I want. It comes from the Unix diff utility, which yields a list of all the differences between two files. More generally it means an unselective and microscopically thorough comparison between two versions of something.

2.2 Listing

Perhaps the most frequently typed command on the Unix command line is **ls**, short for "list" (Listing 2.1).

Listing 2.1: Listing files and directories with **ls**. (Output will vary.)

```
$ ls
Desktop
Downloads
sonnet_1.txt
sonnet_1_reversed.txt
```

The **ls** command simply lists all the files and directories in the current directory (except for those that are *hidden*, which we'll learn more about in a moment). In this sense, it's effectively a command–line version of the graphical browser used to show files and directories (also called "folders"), as seen in Figure 2.3. (We'll sharpen our understanding of directories and folders in Chapter 4.) As with a graphical file browser, the output in Listing 2.1 is just a sample, and results will differ based on the details of your system. (This goes for all the **ls** examples, so don't be concerned if there are minor differences in output.)

The **ls** command can be used to check if a file (or directory) exists, because trying to **ls** a nonexistent file results in an error message, as seen in Listing 2.2.

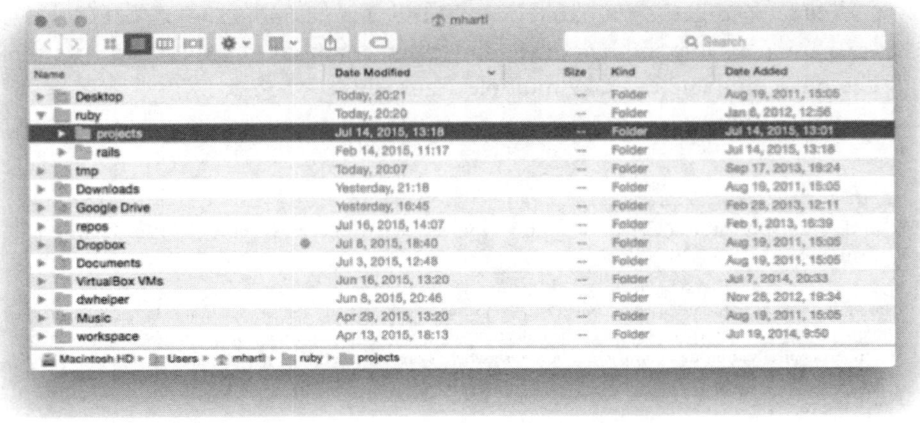

Figure 2.3: The graphical equivalent of **ls**.

Listing 2.2: Running `ls` on a nonexistent file.

```
$ ls foo
ls: foo: No such file or directory
$ touch foo
$ ls foo
foo
```

Listing 2.2 uses the **touch** command to create an empty file with the name **foo** (Box 2.2), so the second time we run `ls` the error message is gone. (The stated purpose of **touch** is to change the modification time on files or directories, but (ab)using **touch** to create empty files as in Listing 2.2 is a common Unix idiom.)

Box 2.2: Foo, Bar, Baz, etc.

When reading about computers, you will encounter certain strange words—words like *foo*, *bar*, and *baz*—with surprising frequency. Indeed, in addition to `ls foo` and `touch foo`, we have already seen three such references in this tutorial: in a typical command-line command (Figure 1.2), when getting out of trouble (`grep foobar` in Box 1.3), and yet again in a man page (Listing 1.4). The first two were my own uses, but the last I had nothing to do with:

```
...if name contains a slash (/) then man interprets
 it as a file specification, so that you can do man
 ./foo.5 or even man /cd/foo/bar.1.gz.
```

Here we see both *foo* and *bar* making an appearance in the man page for man itself—an unambiguous testament to their ubiquity in computing.

What is the origin of these odd terms? As usual, the Jargon File (via its entry on *foo*) enlightens us:

foo: /foo/
1. interj. Term of disgust.
2. [very common] Used very generally as a sample name for absolutely anything, esp. programs and files (esp. scratch files).
3. First on the standard list of metasyntactic variables used in syntax examples. See also bar, baz, qux, quux, etc.
   When "foo" is used in connection with "bar" it has generally traced to the WWII-era Army slang acronym *FUBAR* [see original meaning, https://www.urbandictionary .com/define.php?term=fubar], later modified to *foobar*. Early versions of the Jargon File

interpreted this change as a post-war bowdlerization, but it now seems more likely that FUBAR was itself a derivative of "foo" perhaps influenced by German *furchtbar* (terrible) — "foobar" may actually have been the *original* form.

Following the link to metasyntactic variables (http://www.catb.org/jargon/html/M/metasyntactic-variable.html), we then find the following:

metasyntactic variable: n.

A name used in examples and understood to stand for whatever thing is under discussion, or any random member of a class of things under discussion. The word foo is the canonical example. To avoid confusion, hackers never (well, hardly ever) use "foo" or other words like it as permanent names for anything. In filenames, a common convention is that any filename beginning with a metasyntactic-variable name is a scratch file that may be deleted at any time.

Metasyntactic variables are so called because (1) they are variables in the metalanguage used to talk about programs, etc.; (2) they are variables whose values are often variables (as in usages like "the value of f(foo,bar) is the sum of foo and bar"). However, it has been plausibly suggested that the real reason for the term "metasyntactic variable" is that it sounds good.

In other words, if you want to create a file, and the name doesn't matter, the name is usually "foo". Once you've used "foo", the next file is called "bar", the one after that "baz". Continuations from there vary ("quux" is one common choice), but in many cases three is enough.

A common pattern when using the command line is changing directories using **cd** (covered in Chapter 4) and then immediately typing **ls** to view the contents of the directory. This lets us orient ourselves, and is a good first step toward whatever our next action might be.

One useful ability of **ls** is support for the *wildcard character* ***** (read "star"). For example, to list all files ending in ".txt", we would type this:

```
$ ls *.txt
sonnet_1.txt
sonnet_1_reversed.txt
```

Here ***.txt** (read "star dot tee-ex-tee") automatically expands to all the filenames that match the pattern "any string followed by .txt".

There are three particularly important optional forms of **ls**, starting with the "long form", using the option **-l** (read "dash-ell"):

```
$ ls -l *.txt
total 16
-rw-r--r-- 1 mhartl staff  87 Jul 20 18:05 sonnet_1.txt
-rw-r--r-- 1 mhartl staff 294 Jul 21 12:09 sonnet_1_reversed.txt
```

For now, you can safely ignore most of the information output by **ls -l**, but note that the long form lists a date and time indicating the last time the file was modified. The number before the date is the *size* of the file, in bytes.[4]

A second powerful **ls** variant is "list by **r**eversed **t**ime of modification (**l**ong format)", or **ls -rtl**, which lists the long form of each file or directory in order of how recently it was modified (*reversed* so that the most recently modified entries appear at the bottom of the screen for easy inspection). This is particularly useful when there are a lot of files in the directory but you really only care about seeing the ones that have been modified recently, such as when confirming a file download. We'll see an example of this in Section 3.1, but you are free to try it now:

```
$ ls -rtl
<results system-dependent>
```

By the way, **-rtl** is the commonly used compact form, but you can also pass the options individually, like this:

```
$ ls -r -t -l
```

In addition, their order is irrelevant, so typing **ls -trl** gives the same result.

2.2.1   Hidden Files

Finally, Unix has the concept of "hidden files (and directories)", which don't show up by default when listing files. Hidden files and directories are identified by starting with

---

4. A *bit* is one piece of yes-or-no information (such as a 1 or a 0), and a *byte* is eight bits. Bytes are probably most familiar from "megabytes" and "gigabytes", which represent a million and a billion bytes, respectively. (The official story is a little more complicated, but the level of detail here is certainly enough to be *dangerous*.)

a dot ., and are commonly used for things like storing user preferences. For example, in Part III, we'll create a file called **.gitignore** that tells a particular program (Git) to ignore files matching certain patterns. As a concrete example, to ignore all files ending in ".txt", we could do this:

```
$ echo "*.txt" > .gitignore
$ cat .gitignore
*.txt
```

If we then run **ls**, the file won't show up, because it's hidden:

```
$ ls
sonnet_1.txt
sonnet_1_reversed.txt
```

To get **ls** to display hidden files and directories, we need to pass it the **-a** option (for "all"):

```
$ ls -a
.              .gitignore         sonnet_1_reversed.txt
..             sonnet_1.txt
```

Now **.gitignore** shows up, as expected. (We'll learn what . and .. refer to in Section 4.3.)

2.2.2 Exercises

1. What's the command to list all the non-hidden files and directories that start with the letter "s"?

2. What is the command to list all the non-hidden files that contain the string "onnet", long-form by reverse modification time? *Hint*: Use the wildcard operator at both the beginning and the end.

3. What is the command to list *all* files (including hidden ones) by reverse modification time, in long form?

2.3 Renaming, Copying, Deleting

Next to listing files, probably the most common file operations involve renaming, copying, and deleting them. As with listing files, most modern operating systems provide a graphical user interface to such tasks, but in many contexts it is more convenient to perform them at the command line. *Note*: If you're using macOS, you should follow the instructions in Box 2.3 at this time.

Box 2.3: Switching macOS to Bash

If you're using macOS, at this point you should make sure you're using the right shell program for this tutorial. The default shell as of macOS Catalina is *Z shell* (Zsh), but to get results consistent with this tutorial you should switch to the shell known as *Bash*.

The first step is to determine which shell your system is running, which you can do using the echo command (Section 1.3):

```
$ echo $SHELL
/bin/bash
```

This prints out the $SHELL environment variable. If you see the result shown above, indicating that you're already using Bash, you're done and can proceed with the rest of the tutorial. (In rare cases, $SHELL may differ from the current shell, but the procedure below will still correctly change from one shell to another.) The alert shown in Listing 1.1 is safe to ignore. For more information, including how to switch to and use Z shell with this tutorial, see the Learn Enough blog post "Using Z Shell on Macs with the Learn Enough Tutorials" (https://news.learnenough.com/macos-bash-zshell).

The other possible result of echo is this:

```
$ echo $SHELL
/bin/zsh
```

If that's the result you get, you should use the chsh ("change shell") command as follows:

```
$ chsh -s /bin/bash
```

You'll almost certainly be prompted to type your system password at this point, which you should do. Then completely exit your shell program using Command-Q and relaunch it.

You can confirm that the change succeeded using echo:

```
$ echo $SHELL
/bin/bash
```

At this point, you will probably start seeing the alert shown in Listing 1.1, which you should ignore.

Note that the procedure above is entirely reversible, so there is no need to be concerned about damaging your system. See "Using Z Shell on Macs with the Learn Enough Tutorials" for more information.

The way to rename a file is with the **mv** command, short for "move":

```
$ echo "test text" > test
$ mv test test_file.txt
$ ls
test_file.txt
```

This renames the file called **test** to **test_file.txt**. The final step in the example runs **ls** to confirm that the file renaming was successful, but system-specific files other than the test file are omitted from the output shown. (The name "move" comes from the general use of **mv** to move a file to a different directory (Chapter 4), possibly renaming it en route. When the origin and target directories coincide, such a "move" reduces to a simple renaming.)

The way to copy a file is with **cp**, short for "copy":

```
$ cp test_file.txt second_test.txt
$ ls
second_test.txt
test_file.txt
```

Finally, the command for deleting a file is **rm**, for "remove":

```
$ rm second_test.txt
remove second_test.txt? y
$ ls second_test.txt
ls: second_test.txt: No such file or directory
```

Note that, on many systems, by default you will be prompted to confirm the removal of the file. Any answer starting with the letter "y" or "Y" will cause the file to be deleted, and any other answers will prevent the deletion from occurring.

By the way, in the calls to **cp** and **rm** above, I would almost certainly not type out **test_file.txt** or **second_test.txt**. Instead, I would type something like **test** →| or **sec** →| (where →| represents the tab key (Table 1.1)), thereby making use of *tab completion* (Box 2.4).

Box 2.4: Tab Completion

Most modern command-line programs (shells) support *tab completion*, which involves automatically completing a word if there's only one valid match on the system. For example, if the only file starting with the letters "tes" is test_file, we could create the command to remove it as follows:

```
$ rm tes→|
```

where →| is the tab key (Table 1.1). The shell would then complete the filename, yielding rm test_file. Especially with longer filenames (or directories), tab completion can save a huge amount of typing. It also lowers the *cognitive load*, since it means you don't have to remember the full name of the file—only its first few letters.

If the match is ambiguous, as would happen if we had files called foobarquux and foobazquux, the word will be completed only as far as possible, so

```
$ ls foo→|
```

would be completed to

```
$ ls fooba
```

If we then hit tab *again*, we would see a list of matches:

```
$ ls fooba→|
foobarquux foobazquux
```

We could then type more letters to resolve the ambiguity, so typing the r after fooba and hitting →| would yield

```
$ ls foobar→|
```

which would be completed to foobarquux. This situation is common enough that experienced command-line users will often just hit something like f →|→| to get the shell to show all the possibilities:

```
$ ls f →|→|
figure_1.png foobarquux  foobazquux
```

Additional letters would then be typed as usual to resolve the ambiguity.

The default behavior of **rm** on an unconfigured Unix system is actually to remove the file without confirmation, but (because deletion is irreversible) many systems *alias* the **rm** command to use an option to turn on confirmation. (As you can verify by running **man rm**, this option is **-i**, so in fact **rm** is really **rm -i**.) There are many situations where confirmation is inconvenient, though, such as when you're deleting a list of files and don't want to have to confirm each one. This is especially common when using the wildcard * introduced in Section 2.2. For example, to remove all the files ending with ".txt" using a single command, *without* having to confirm each one, you can type this:

```
$ rm -f *.txt
```

Here **-f** (for "force") overrides the implicit **-i** option and removes all files immediately. (N.B. You are now in a position to understand the command in Figure 1.2.)

2.3.1 Unix Terseness

One thing you might notice is that the commands in this section and in Section 2.2 are short: Instead of **list**, **move**, **copy**, and **remove**, we have **ls**, **mv**, **cp**, and **rm**. Because the former command names are easier to understand and memorize, you may wonder why the actual commands aren't longer (Figure 2.4).

The answer is that Unix dates from a time when most computer users logged on to centralized servers over slow connections, and there could be a noticeable delay between the time users pressed a key and the time it appeared on the terminal. For frequently used commands like listing files, the difference between **list** and **ls** or **remove** and **rm** could be significant. As a result, the most commonly used Unix commands tend to be only two or three letters long. Because it makes them more difficult

Figure 2.4: The terseness of Unix commands can be a source of confusion.

to memorize, this can be a minor inconvenience when learning them, but over a lifetime of command-line use the savings represented by, say, **mv** really add up.

2.3.2 Exercises

1. Use the **echo** command and the redirect operator **>** to make a file called **foo.txt** containing the text "hello, world". Then, using the **cp** command, make a copy of **foo.txt** called **bar.txt**. Using the **diff** command, confirm that the contents of both files are the same.

2. By combining the **cat** command and the redirect operator **>**, create a copy of **foo.txt** called **baz.txt** *without* using the **cp** command.

3. Create a file called **quux.txt** containing the contents of **foo.txt** followed by the contents of **bar.txt**. *Hint*: As noted in Section 2.1.1, **cat** can take multiple arguments.

4. How do **rm nonexistent** and **rm -f nonexistent** differ for a nonexistent file?

2.4 Summary

Important commands from this chapter are summarized in Table 2.1.

Table 2.1: Important commands from Chapter 2.

Command	Description	Example
`>`	Redirect output to filename	`$ echo foo > foo.txt`
`>>`	Append output to filename	`$ echo bar >> foo.txt`
`cat <file>`	Print contents of file to screen	`$ cat hello.txt`
`diff <f1> <f2>`	Diff files 1 & 2	`$ diff foo.txt bar.txt`
`ls`	List directory or file	`$ ls hello.txt`
`ls -l`	List long form	`$ ls -l hello.txt`
`ls -rtl`	Long by reverse modification time	`$ ls -rtl`
`ls -a`	List all (including hidden)	`$ ls -a`
`touch <file>`	Create an empty file	`$ touch foo`
`mv <old> <new>`	Rename (move) from old to new	`$ mv foo bar`
`cp <old> <new>`	Copy old to new	`$ cp foo bar`
`rm <file>`	Remove (delete) file	`$ rm foo`
`rm -f <file>`	Force-remove file	`$ rm -f bar`

2.4.1 Exercises

1. By copying and pasting the text from the HTML version (https://www
.learnenough.com/sonnet) of Figure 2.5, use **echo** to make a file called
sonnet_1_ complete.txt containing the full (original) text of Shakespeare's
first sonnet. *Hint*: You may recall getting stuck when **echo** was followed by an
unmatched double quote (Section 1.3 and Box 1.3), as in **echo "**, but in fact this
construction allows you to print out a multi–line block of text. Just remember to
put a closing quote at the end, and then redirect to a file with the appropriate
name. Check that the contents are correct using **cat** (Figure 2.2).

2. Type the sequence of commands needed to create an empty file called **foo**, rename
it to **bar**, and copy it to **baz**.

3. What is the command to list only the files starting with the letter "b"? *Hint*: Use
a wildcard.

FRom fairest creatures we desire increase,
That thereby beauties *Rose* might neuer die,
But as the riper should by time decease,
His tender heire might beare his memory:
But thou contracted to thine owne bright eyes,
Feed'st thy lights flame with selfe substantiall fewell,
Making a famine where aboundance lies,
Thy selfe thy foe,to thy sweet selfe too cruell:
Thou that art now the worlds fresh ornament,
And only herauld to the gaudy spring,
Within thine owne bud buriest thy content,
And tender chorle makst wast in niggarding:
Pitty the world,or else this glutton be,
To eate the worlds due,by the graue and thee.

Figure 2.5: Shakespeare's first sonnet (copy-and-pastable version is available at https://www.learnenough.com/sonnet).

4. Remove both **bar** and **baz** using a *single* call to **rm**. *Hint*: If those are the only two files in the current directory that start with the letter "b", you can use the wildcard pattern from the previous exercise.

CHAPTER 3

Inspecting Files

Having seen how to create and manipulate files, now it's time to learn how to examine their contents. This is especially important for files too long to fit on a single screen. In particular, we saw starting in Section 2.1 how to use the **cat** command to dump the file contents to the screen, but this doesn't work very well for longer files.

3.1 Downloading a File

To give us a place to start, rather than creating a long file by hand (which is cumbersome) we'll download a file from the Internet using the powerful **curl** utility. Sometimes written as "cURL", the **curl** program allows us to interact with a URL[1] at the command line. Although it's not part of the core Unix command set, the **curl** command is widely available for installation on Unix systems. To make sure it's available on your system, we can use the **which** command, which looks to see if the given program is available at the command line.[2] The way to use it is to type **which** followed by the name of the program—in this case, **curl**:

```
$ which curl
/usr/bin/curl
```

I've shown the output on my system (**/usr/bin/curl**, usually read as "user bin curl"), but the result on your system may differ. In particular, if the result is just a blank line, you will have to install **curl**, which you can do by Googling for "install curl" followed

---

1. URL is short for Uniform Resource Locator, and in practice usually just means "web address".

2. Technically, **which** locates a file on the user's *path*, which is a list of directories where executable programs are located.

by the name of your operating system (e.g., "install curl macos"). (This sort of "Google for it" installation step is classic technical sophistication (Box 1.4).)

Once **curl** is installed, we can use the command in Listing 3.1 to download a file called **sonnets.txt**, which contains a large corpus of text.[3]

Listing 3.1: Using **curl** to download a longer file.

```
$ curl -OL https://cdn.learnenough.com/sonnets.txt
$ ls -rtl
```

Be sure to copy the command exactly; in particular, note that the option **-OL** contains a capital letter "O" (**O**) and not a zero (**0**). (Figuring out what these options do is left as an exercise (Section 3.5.1).) Also, on some systems (for mysterious reasons) you might have to run the command twice to get it to work; by inspecting the results of **ls -rtl**, you should be able to tell if the initial call to **curl** created the file **sonnets.txt** as expected. (If you do have to repeat the **curl** command, you could press up arrow twice to retrieve it, but see Box 3.1 for alternatives.)

The result of running Listing 3.1 is **sonnets.txt**, a file containing all 154 of Shakespeare's sonnets. This file contains 2620 lines, far too many to fit on one screen. Learning how to inspect its contents is the goal of the rest of this chapter. (Among other things, we'll learn how to determine that it has 2620 lines without counting them all by hand.)

Box 3.1: Repeating Previous Commands

Repeating previous commands is a frequent task when using the command line. So far in this tutorial, we've used the up-arrow key to retrieve (and possibly edit) previous commands, but this isn't the only possibility. An even quicker way to find and immediately run a previous command involves using the exclamation point !, which in the context of software development is usually pronounced "bang". To run the previous command exactly as written, we can use "bang bang":

---

3. If for any reason using **curl** fails, you can always visit the URL in a browser and then use the **File > Save As** feature to save it to your local disk.

```
$ echo "foo"
foo
$ !!
echo "foo"
foo
```

A closely related usage is "bang" followed by some number of characters, which runs the last command that started with those characters. For example, to run the last `curl` command, we could type this:

```
$ !curl
```

This would save us the trouble of typing out the options, the URL, etc. Depending on our history of commands, the even terser `!cu` or `!c` would work as well. This technique is especially useful when the desired command last happened many commands ago, which can make hitting up arrow cumbersome.

A second and incredibly powerful technique is ^R, which lets you search interactively through your previous commands, and then optionally edit the result before executing. For example, we could try this to bring up the last `curl` command:

```
$ ^R
(reverse-i-search)`': curl
```

On most systems, hitting return would then put the last `curl` command after our prompt and allow us to edit it (if desired) before hitting return to execute it. When your workflow happens to involve repeatedly running a variety of similar commands, sometimes it can seem like "all commands start with ^R".

3.1.1 Exercises

1. Use the command **curl -I https://www.learnenough.com/** to fetch the *HTTP header* for the Learn Enough website. What is the HTTP status code for the address? How does this differ from the status code for `learnenough.com` (without the **https://**)?

2. Using **ls**, confirm that **sonnets.txt** exists on your system. How big is it in bytes? *Hint*: Recall from Section 2.2 that the "long form" of **ls** displays a byte count.

3. The byte count in the previous exercise is high enough that it's more naturally thought of in *kilobytes* (often treated as 1000 bytes, but actually equal to

$2^{10} = 1024$ bytes). By adding the **-h** ("human-readable") option to **ls**, list the long form of the sonnets file with a human-readable byte count.

4. Suppose you wanted to list the files and directories using **h**uman-readable byte counts, **a**ll, by **r**everse **t**ime-sorted **l**ong-form. What command would you use? Why might this command be a personal favorite of the author of this tutorial?[4]

3.2 Making Heads and Tails of It

Two complementary commands for inspecting files are **head** and **tail**, which respectively allow us to view the beginning (head) and end (tail) of the file. The **head** command shows the first 10 lines of the file (Listing 3.2).

Listing 3.2: Looking at the head of the sample text file.

```
$ head sonnets.txt
Shake-speare's Sonnets

I

From fairest creatures we desire increase,
That thereby beauty's Rose might never die,
But as the riper should by time decease,
His tender heir might bear his memory:
But thou contracted to thine own bright eyes,
Feed'st thy light's flame with self-substantial fuel
```

Similarly, **tail** shows the last 10 lines of the file (Listing 3.3).

Listing 3.3: Looking at the tail of the sample text file.

```
$ tail sonnets.txt
The fairest votary took up that fire
Which many legions of true hearts had warm'd;
And so the general of hot desire
Was, sleeping, by a virgin hand disarm'd.
This brand she quenched in a cool well by,
Which from Love's fire took heat perpetual,
```

---

4. Having known about **ls -a** and **ls -rtl** for a while—which together yield the suggestive command **ls -artl**—one day I decided to add an "h" (for obvious reasons [https://www.michaelhartl.com/]). This is actually how I accidentally discovered the useful **-h** option some years ago.

```
Growing a bath and healthful remedy,
For men diseas'd; but I, my mistress' thrall,
 Came there for cure and this by that I prove,
 Love's fire heats water, water cools not love.
```

These two commands are useful when (as is often the case) you know for sure you only need to inspect the beginning or end of a file.

3.2.1 Wordcount and Pipes

By the way, I didn't recall offhand how many lines **head** and **tail** show by default. Since there are only 10 lines in the output, I could have counted them by hand, but in fact I was able to figure it out using the **wc** command (short for "wordcount"; recall Figure 2.4).

The most common use of **wc** is on full files. For example, we can run **sonnets.txt** through **wc**:

```
$ wc sonnets.txt
  2620  17670  95635 sonnets.txt
```

Here the three numbers indicate how many lines, words, and bytes there are in the file, so there are 2620 lines (thereby fulfilling the promise made at the end of Section 3.1), 17670 words, and 95635 bytes.

You are now in a position to be able to guess one method for determining how many lines are in **head sonnets.txt**. In particular, we can combine **head** with the redirect operator (Section 2.1) to make a file with the relevant contents, and then run **wc** on it, as shown in Listing 3.4.

Listing 3.4: Redirecting **head** and running **wc** on the result.

```
$ head sonnets.txt > sonnets_head.txt
$ wc sonnets_head.txt
  10   46   294 sonnets_head.txt
```

We see from Listing 3.4 that there are 10 lines in **head wc** (along with 46 words and 294 bytes). The same method, of course, would work for **tail**.

On the other hand, you might get the feeling that it's a little unclean to make an intermediate file just to run **wc** on it, and indeed there's a way to avoid it using a technique called *pipes*. Listing 3.5 shows how to do it.

Listing 3.5: Piping the result of **head** through **wc**.

```
$ head sonnets.txt | wc
   10    46   294
```

The command in Listing 3.5 runs **head sonnets.txt** and then *pipes* the result through **wc** using the pipe symbol **|** (Shift-backslash on most QWERTY keyboards). The reason this works is that the **wc** command, in addition to taking a filename as an argument, can (like many Unix programs) take input from "standard in" (compare to "standard out" mentioned in Section 1.3), which in this case is the output of **head sonnets.txt** shown in Listing 3.2. The **wc** program takes this input and counts it the same way it counts a file, yielding the same line, word, and byte counts as Listing 3.4.

3.2.2 Exercises

1. By piping the results of **tail sonnets.txt** through **wc**, confirm that (like **head**) the **tail** command outputs 10 lines by default.

2. By running **man head**, learn how to look at the first **n** lines of the file. By experimenting with different values of **n**, find a **head** command to print out just enough lines to display the first sonnet in its entirety (Figure 1.11).

3. Pipe the results of the previous exercise through **tail** (with the appropriate options) to print out *only* the 14 lines composing Sonnet 1. *Hint*: The command will look something like **head -n <i> sonnets.txt | tail -n <j>**, where **<i>** and **<j>** represent the numerical arguments to the **-n** option.

4. One of the most useful applications of **tail** is running **tail -f** to view a file that's actively changing. This is especially common when monitoring files used to log the activity of, e.g., web servers, a practice known as "tailing the log file". To simulate the creation of a log file, run **ping learnenough.com > learnenough.log** in one terminal tab. (The **ping** command "pings" a server to see if it's working.) In a second tab, type the command to tail the log file. (At this point, both tabs will be stuck, so once you've gotten the gist of **tail -f** you should use the technique from Box 1.3 to get out of trouble.)

3.3 Less Is More

Unix provides two utilities for the common task of wanting to look at more than just the head or tail of a file. The older of these programs is called **more**, but (I'd guess initially as a tongue-in-cheek joke) there's a more powerful variant called **less**.[5] The **less** program is interactive, so it's hard to capture in print, but here's roughly what it looks like:

```
$ less sonnets.txt
Shake-speare's Sonnets

I

From fairest creatures we desire increase,
That thereby beauty's Rose might never die,
But as the riper should by time decease,
His tender heir might bear his memory:
But thou contracted to thine own bright eyes,
Feed'st thy light's flame with self-substantial fuel,
Making a famine where abundance lies,
Thy self thy foe, to thy sweet self too cruel:
Thou that art now the world's fresh ornament,
And only herald to the gaudy spring,
Within thine own bud buriest thy content,
And tender churl mak'st waste in niggarding:
 Pity the world, or else this glutton be,
 To eat the world's due, by the grave and thee.

II

When forty winters shall besiege thy brow,
And dig deep trenches in thy beauty's field,
sonnets.txt
```

The point of **less** is that it lets you navigate through the file in several useful ways, such as moving one line up or down with the arrow keys, pressing the spacebar to move a page down, and pressing **^F** to move forward a page (i.e., the same as spacebar) or **^B** to move back a page. To quit **less**, type **q** (for "quit").

Perhaps the most powerful aspect of **less** is the forward slash key **/**, which lets you search through the file from beginning to end. For example, suppose we wanted to search through **sonnets.txt** for "rose" (Figure 3.1),[6] one of the most frequently

---

5. On some systems, apparently they're exactly the same program, so **less** really is **more** (or, more accurately, **more** is **less**).

6. Image courtesy of Shuang Li/Shutterstock.

Figure 3.1: A famous rose from the time of Shakespeare.

used images in the *Sonnets*.[7] The way to do this in **less** is to type **/rose** (read "slash rose"), as shown in Listing 3.6.

Listing 3.6: Searching for the string "rose" using **less**.

```
Shake-speare's Sonnets

I

From fairest creatures we desire increase,
That thereby beauty's Rose might never die,
But as the riper should by time decease,
His tender heir might bear his memory:
But thou contracted to thine own bright eyes,
Feed'st thy light's flame with self-substantial fuel,
Making a famine where abundance lies,
Thy self thy foe, to thy sweet self too cruel:
Thou that art now the world's fresh ornament,
And only herald to the gaudy spring,
Within thine own bud buriest thy content,
And tender churl mak'st waste in niggarding:
 Pity the world, or else this glutton be,
 To eat the world's due, by the grave and thee.

II

When forty winters shall besiege thy brow,
And dig deep trenches in thy beauty's field,
/rose
```

The result of pressing return after typing **/rose** in Listing 3.6 is to highlight the first occurrence of "rose" in the file. You can then press **n** to navigate to the next match, or **N** to navigate to the previous match.

---

7. Although Shakespeare's sonnets are undated, most of them were probably composed during the reign of Queen Elizabeth, whose royal house adopted a rose (Figure 3.1) as its heraldic emblem. Given this context, Shakespeare's choice of floral imagery isn't surprising, but in fact only a few commentators on the *Sonnets* have noticed the seemingly obvious reference.

The last two essential **less** commands are **G** to move to the end of the file and **1G** (that's **1** followed by **G**) to move back to the beginning. Table 3.1 summarizes what are in my view the most important key combinations (i.e., the ones I think you need to be *dangerous*), but if you're curious you can find a longer list of commands at the Wikipedia page on less.

I encourage you to get in the habit of using **less** as your go-to utility for looking at the contents of a file. The skills you develop have other applications as well; for example, the man pages (Section 1.4) use the same interface as **less**, so by learning about **less** you'll get better at navigating the man pages as well.

Table 3.1: The most important **less** commands.

Command	Description	Example
up & down arrow keys	Move up or down one line	
spacebar	Move forward one page	
^F	Move forward one page	
^B	Move back one page	
G	Move to end of file	
1G	Move to beginning of file	
/<string>	Search file for string	/rose
n	Move to next search result	
N	Move to previous search result	
q	Quit **less**	

3.3.1 Exercises

1. Run **less** on **sonnets.txt**. Go down three pages and then back up three pages. Go to the end of the file, then to the beginning, then quit.

2. Search for the string "All" (case-sensitive). Go forward a few occurrences, then back a few occurrences. Then go to the beginning of the file and count the occurrences by searching forward until you hit the end. Compare your count to the result of running **grep All sonnets.txt | wc**. (We'll learn about **grep** in Section 3.4.)

3. Using **less** and **/** ("slash"), find the sonnet that begins with the line "Let me not". Are there any other occurrences of this string in the *Sonnets*? *Hint*:

Press **n** to find the next occurrence (if any). *Extra credit*: Listen to the sonnet (https://www.youtube.com/watch?v=bt7OynPUIY8) in both modern and original pronunciation. Which version's rhyme scheme is better?

4. Because **man** uses **less**, we are now in a position to search man pages interactively. By searching for the string "sort" in the man page for **ls**, discover the option to sort files by size. What is the command to display the long form of files sorted so the largest files appear at the bottom? *Hint*: Use **ls -rtl** as a model.

3.4 Grepping

One of the most powerful tools for inspecting file contents is **grep**, which probably stands for something, but it's not important what. (We'll actually mention it in a moment.) Indeed, *grep* is frequently used as a verb, as in "You should totally grep that file."

The most common use of **grep** is just to search for a substring in a file. For example, we saw in Section 3.3 how to use **less** to search for the string "rose" in Shakespeare's sonnets. Using **grep**, we can find the references directly, as shown in Listing 3.7.

Listing 3.7: Finding the occurrences of "rose" in Shakespeare's sonnets.

```
$ grep rose sonnets.txt
The rose looks fair, but fairer we it deem
As the perfumed tincture of the roses.
Die to themselves. Sweet roses do not so;
Roses of shadow, since his rose is true?
Which, like a canker in the fragrant rose,
Nor praise the deep vermilion in the rose;
The roses fearfully on thorns did stand,
 Save thou, my rose, in it thou art my all.
I have seen roses damask'd, red and white,
But no such roses see I in her cheeks;
```

With the command in Listing 3.7, it appears that we are in a position to count the number of lines containing references to the word "rose" by piping to **wc** (as in Section 3.3), as shown in Listing 3.8.

Listing 3.8: Piping the results of **grep** to **wc**.

```
$ grep rose sonnets.txt | wc
   10    82    419
```

Listing 3.8 tells us that 10 lines contain "rose" (or "roses", since "rose" is a substring of "roses"). But you may recall from Figure 1.11 that Shakespeare's first sonnet contains "Rose" with a *capital* "R". Referring to Listing 3.7, we see that this line has in fact been missed. This is because **grep** is case-sensitive by default, and "rose" doesn't match "Rose".

As you might suspect, **grep** has an option to perform case-insensitive matching as well. One way to figure it out is to search through the **man** page for **grep**:

- Type **man grep**.
- Type **/case** and then return.
- Read off the result (Figure 3.2).

(As noted briefly in Section 1.4, the man pages use the same interface as the **less** command we met in Section 3.3, so we can search through them using /.)

Applying the result of the above procedure yields Listing 3.9. Comparing the results of Listing 3.9 with Listing 3.8, we see that we now have 12 matching lines instead of only 10, so there must be a total of $12 - 10 = 2$ lines containing "Rose" (but not "rose") in the *Sonnets*.[8]

Listing 3.9: Doing a case-insensitive grep.

```
$ grep -i rose sonnets.txt | wc
   12    96    508
```

The **grep** utility gets its name from a pattern-matching system called *regular expressions* (also called *regexes* for short): *grep* stands for "**g**lobally search a **r**egular **e**xpression and **p**rint." A full treatment of regular expressions is well beyond the scope of this tutorial, but before moving on we'll sample just a small taste.

---

8. Actually, "ROSE", "RoSE", "rOSE", etc., all match as well, but "Rose" is the likeliest candidate. Confirming this hunch is left as an exercise (Section 3.4.1).

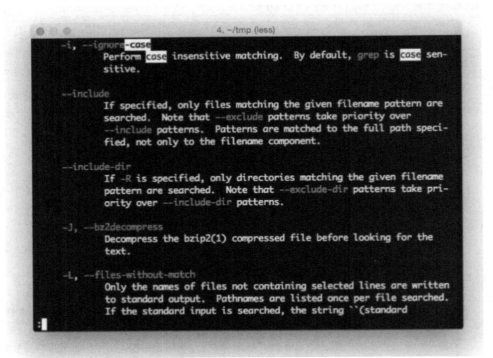

Figure 3.2: The result of searching **man grep** for "case".

As one simple example, let's match every line in **sonnets.txt** that has a word beginning with the letters "ro", followed by any number of (lowercase) letters, and ending in "s". The way to represent "any letter" with a regular expression is **[a-z]**, and following a pattern with an asterisk ***** matches "zero or more" of that thing. Thus, **ro[a-z]*s** matches "ro" and "s" with zero or more letters in between. We can add spaces to the beginning and end to ensure that the match consists of entire words, like this:

```
$ grep ' ro[a-z]*s ' sonnets.txt
  To that sweet thief which sourly robs from me.
Die to themselves. Sweet roses do not so;
When rocks impregnable are not so stout,
He robs thee of, and pays it thee again.
```

```
The roses fearfully on thorns did stand,
I have seen roses damask'd, red and white,
But no such roses see I in her cheeks;
```

We can see that the regular expression matches strings like "robs" and "rocks" in addition to "roses".

In general, one of the best tools for learning how to use regexes is an *online regex builder*, such as regex101, which lets you build up regexes interactively (Figure 3.3). Unfortunately, **grep** often doesn't support the precise format used by regex builders (including hard-to-guess requirements for "escaping out" special characters), and precision in regular expressions is everything. As a result, despite its name origins, the truth is I rarely use the regular expression capabilities of **grep**. By the time the situation calls for regexes, I'm far likelier to reach for a text editor (Part II) or a full-strength programming language (*Learn Enough JavaScript to Be Dangerous* (https://www.learnenough.com/javascript), *Learn Enough Ruby to Be Dangerous* (https://www.learnenough.com/ruby)).

Nevertheless, the aspects of **grep** discussed in this section are nearly enough to be *dangerous*, covering a huge number of common cases (including the important application of *grepping processes* (Box 3.2)). We'll see one final **grep** variant in Chapter 4 as part of our discussion of Unix directories.

Box 3.2: Grepping Processes

One of the many uses of `grep` is filtering the Unix *process list* for running programs that match a particular string. (On Unix-like systems such as Linux and macOS, user and system tasks each take place within a well-defined container called a *process*.) This is especially useful when there's a rogue process on your system that needs to be killed. (A good way to find such processes is by running the `top` command, which shows the processes consuming the most resources.)

For example, at one point in the *Ruby on Rails Tutorial* book (https://www.railstutorial.org/book), it's important to eliminate a program called `spring` from the process list. To do this, first the processes need to be found, and the way to see all the processes on your system is to use the `ps` command with the `aux` options:

```
$ ps aux
```

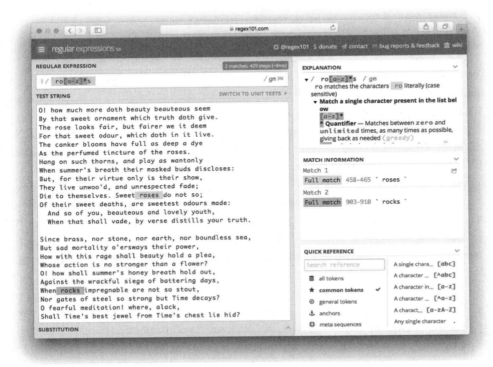

Figure 3.3: An online regex builder (https://regex101.com/).

Per the usual practice of Unix terseness (Figure 2.4), ps is short for "process status". And for confusing and obscure reasons, options to ps aren't written with a dash (so it's ps aux instead of ps -aux). (How on earth are you supposed to know this? That's what this tutorial is for.)

To filter the processes by program name, you pipe the results of ps through grep:

```
$ ps aux | grep spring
ubuntu 12241 0.3 0.5 589960 178416 ? Ssl Sep20 1:46
spring app | sample_app | started 7 hours ago
```

The result shown gives some details about the process, but the most important thing is the first number, which is the *process id*, or pid (often pronounced to rhyme with "kid"). To eliminate an unwanted process, we use the `kill` command to issue the Unix terminate code (which happens to be 15) to the pid:

```
$ kill -15 12241
```

This is the technique I recommend for killing individual processes, such as a rogue web server (with the pid found via `ps aux | grep server`), but sometimes it's convenient to kill all the processes matching a particular process name, such as when you want to kill all the `spring` processes gunking up your system. In this case, you can kill all the processes with name `spring` using the `pkill` command as follows:

```
$ pkill -15 -f spring
```

Any time something isn't behaving as expected, or a process appears to be frozen, it's a good idea to run `top` or `ps aux` to see what's going on, pipe `ps aux` through `grep` to select the suspected processes, and then run `kill -15 <pid>` or `pkill -15 -f <name>` to clear things up.

3.4.1 Exercises

1. By searching **man grep** for "line number", construct a command to find the line numbers in **sonnets.txt** where the string "rose" appears.

2. You should find that the last occurrence of "rose" is (via "roses") on line 2203. Figure out how to go directly to this line when running **less sonnets.txt**. *Hint*: Recall from Table 3.1 that **1G** goes to the top of the file, i.e., line 1. Similarly, **17G** goes to line 17. Etc.

3. By piping the output of **grep** to **head**, print out the first (and *only* the first) line in **sonnets.txt** containing "rose". *Hint*: Use the result of the second exercise in Section 3.2.2.

4. In Listing 3.9, we saw two additional lines that case-insensitively matched "rose". Execute a command confirming that both of the lines contain the string "Rose" (and not, e.g., "rOSe"). *Hint*: Use a case-sensitive **grep** for "Rose".

5. You should find in the previous exercise that there are *three* lines matching "Rose" instead of the two you might have expected from Listing 3.9. This is because there is one line that contains both "Rose" *and* "rose", and thus shows up in both **grep rose** and **grep -i rose**. Write a command confirming that the number of lines

matching "Rose" but *not* matching "rose" is equal to the expected 2. *Hint*: Pipe the result of **grep** to **grep -v**, and then pipe that result to **wc**. (What does **-v** do? Read the man page for **grep** (Box 1.4).)

3.5 Summary

Important commands from this chapter are summarized in Table 3.2.

Table 3.2: Important commands from Chapter 3.

Command	Description	Example
`curl`	Interact with URLs	`$ curl -O https://ex.co`
`which`	Locate a program on the path	`$ which curl`
`head <file>`	Display first part of file	`$ head foo`
`tail <file>`	Display last part of file	`$ tail bar`
`wc <file>`	Count lines, words, bytes	`$ wc foo`
`cmd1 | cmd2`	Pipe `cmd1` to `cmd2`	`$ head foo | wc`
`ping <url>`	Ping a server URL	`$ ping google.com`
`less <file>`	View file contents interactively	`$ less foo`
`grep <string> <file>`	Find string in file	`$ grep foo bar.txt`
`grep -i <string> <file>`	Find case-insensitively	`$ grep -i foo bar.txt`
`ps`	Show processes	`$ ps aux`
`top`	Show processes (sorted)	`$ top`
`kill -<level> <pid>`	Kill a process	`$ kill -15 24601`
`pkill -<level> -f <name>`	Kill matching processes	`$ pkill -15 -f spring`

3.5.1 Exercises

1. The **history** command prints the history of commands in a particular terminal shell (subject to some limit, which is typically large). Pipe **history** to **less** to examine your command history. What was your 17th command?

2. By piping the output of **history** to **wc**, count how many commands you've executed so far.

3. One use of **history** is to grep your commands to find useful ones you've used before, with each command preceded by the corresponding number in the command history. By piping the output of **history** to **grep**, determine the number for the last occurrence of **curl**.

4. In Box 3.1, we learned about **!!** ("bang bang") to execute the previous command. Similarly, **!n** executes command number **n**, so that, e.g., **!17** executes the 17th command in the command history. Use the result from the previous exercise to rerun the last occurrence of **curl**.

5. What do the **O** and **L** options in Listing 3.1 mean? *Hint*: Pipe the output of **curl -h** to **less** and search first for the string **-O** and then for the string **-L**.

CHAPTER 4

Directories

Having examined many of the Unix utilities for dealing with files, the time has come to learn about *directories*, sometimes known by the synonym *folders* (Figure 4.1). As we'll see, many of the ideas developed in the context of files also apply to directories, but there are many differences as well.

4.1 Directory Structure

The structure of Unix-style directories is typically indicated using a list of directory names separated by forward slashes, which we can combine with the **ls** command (Section 2.2) like this:

```
$ ls /Users/mhartl/ruby
```

or like this:

```
$ ls /usr/local/bin
```

As seen in Figure 4.1, these representations correspond to directories in a hierarchical filesystem, with (say) **mhartl** a subdirectory of **Users** and **ruby** a subdirectory of **mhartl**.

Conventions vary when speaking about directories: a *user directory* like **/Users/mhartl** would probably be read as "slash users slash mhartl" or "slash users mhartl", whereas omitting the initial slash in spoken language is common for *system directories*

Figure 4.1: The correspondence between folders & directories.

such as **/usr/local/bin**, which would probably be pronounced "user local bin".[1] Because *all* Unix directories are ultimately subdirectories of the *root directory* **/** (pronounced "slash"), the leading and separator slashes are implied. *Note*: Referring to forward slashes incorrectly as "backslashes" is a source of intense suffering, and should be strictly avoided.

The most important directory for a particular user is the *home directory*, which on my macOS system is **/Users/mhartl**, corresponding to my username (**mhartl**). The home directory can be specified as an *absolute path*, as in **/Users/mhartl**, or using the shorthand for the home directory, the tilde character **~** (which is typed using Shift-backtick, located just to the left of the number 1 on most keyboards). As a result, the two paths shown in Figure 4.1 are identical: **/Users/mhartl/ruby/projects** is the same as **~/ruby/projects**. (Amusingly, the reason the tilde character is used for the home directory is simply because the "Home" key was the same as the key for producing "~" on some early keyboards.)

In addition to user directories, every Unix system has *system directories* for the programs essential for the computer's normal operation. Modifying system files or directories requires special powers granted only to the *superuser*, known as **root**. (This

---

1. For more on Unix system directories, see "What is /usr/local/bin?" (https://unix.stackexchange.com /questions/4186/what-is-usr-local-bin) at the Unix StackExchange. Thanks to reader Joost van der Linden for the suggestion.

use of "root" is unrelated to the "root directory" mentioned above.) The superuser is so powerful that it's considered bad form to log in as **root**; instead, tasks executed as **root** should typically use the **sudo** command (Box 4.1).

Box 4.1: "sudo **make me a sandwich.**"

sudo gives ordinary users the power to execute commands as the superuser. For example, let's try touching a file in the system directory /opt as follows:

```
$ touch /opt/foo
touch: /opt/foo: Permission denied
```

Because normal users don't have permission to modify /opt, the command fails, but it succeeds with sudo:

```
$ sudo touch /opt/foo
Password:
```

As shown, after entering sudo we are prompted to enter our user password; if entered correctly, and if the user has been configured to have sudo privileges (which is the default on most desktop Unix systems), then the command will succeed. As shown in the xkcd comic strip "Sandwich" (https://m.xkcd.com/149/), this pattern of being denied at first, only to succeed using sudo, is a common pattern when using the command line.

To check that the file really was created, we can ls it:

```
$ ls -l /opt/foo
-rw-r--r-- 1 root wheel 0 Jul 23 19:13 /opt/foo
```

Note that (1) a normal user can ls a file in a system directory (without sudo) and (2) the name root appears in the listing, indicating that the superuser owns the file. (The meaning of the second term, wheel, is a little obscure, but you can learn about it on a site called, appropriately enough, superuser (https://superuser.com/questions/191955/what-is-the-wheel-user-in-os-x).)

To remove the file we just created, we again need superuser status:

```
$ rm -f /opt/foo
rm: /opt/foo: Permission denied
$ sudo !!
$ !ls
ls: /opt/foo: No such file or directory
```

Here the first rm fails, so we've run sudo !!, which runs sudo and then the previous command, and we've followed that up with !ls, which runs the previous ls command (Box 3.1).

It's also worth noting the English pronunciation of something like sudo !!, which is important when communicating via spoken language. As noted in Box 3.1, !! is pronounced "bang bang". sudo, meanwhile, is pronounced either "SOO-doo" or "SOO-doh". Both pronunciations are common, though I prefer the former because the do in sudo is in fact just the English word "do". Thus, my preferred pronunciation for sudo !! is "SOO-doo bang bang".

By the way, the su in sudo originally stood for "super-user" (https://pthree.org/2009/12/31/the-meaning-of-su/), but over time its use expanded, and now is usually thought of as "substitute user". sudo is therefore a contraction of "substitute user do", with the substitute user being the superuser by default. Because the superuser can do anything, the command "sudo make me a sandwich" in "Sandwich" succeeds when a mere "make me a sandwich" does not.

4.1.1 Exercises

1. Write in words how you might speak the directory **~/foo/bar**.

2. In **/Users/bill/sonnets**, what is the home directory? What is the username? Which directory is deepest in the hierarchy?

3. For a user with username **bill**, how do **/Users/bill/sonnets** and **~/sonnets** differ (if at all)?

4.2 Making Directories

So far in this tutorial, we've created (and removed) a large number of text files. The time has finally come to make a directory to contain them. Although most modern operating systems include a graphical interface for this task, the Unix way to do it is with **mkdir** (short for "make directory", per Figure 2.4):

```
$ mkdir text_files
```

Having made the directory, we can move the text files there using a wildcard:

```
$ mv *.txt text_files/
```

We can confirm the move worked by listing the directory:

```
$ ls text_files/
sonnet_1.txt        sonnet_1_reversed.txt sonnets.txt
```

(Depending on how closely you've followed this tutorial, your results may vary.)

By default, running **ls** on a directory shows its *contents*, but we can show just the directory using the **-d** option:

```
$ ls -d text_files/
text_files/
```

This usage is especially common with the **-l** option (Section 2.2):

```
$ ls -ld text_files/
drwxr-xr-x 7 mhartl staff 238 Jul 24 18:07 text_files
```

Finally, we can change directories using **cd**:

```
$ cd text_files/
```

Note that **cd** typically supports tab completion, so (as described in Box 2.4) we can actually type **cd tex→|**.

After running **cd**, we can confirm that we're in the correct directory using the "print working directory" command, **pwd**, together with another call to **ls**:

```
$ pwd
/Users/mhartl/text_files
$ ls
sonnet_1.txt        sonnet_1_reversed.txt sonnets.txt
```

These last steps of typing **pwd** to double-check the directory, and especially running **ls** to inspect the directory contents, are a matter of habit for many a grizzled command-line veteran. (Your result for **pwd** will, of course, be different, unless you happen to be using the username "mhartl" on macOS.)

4.2.1 Exercises

1. What is the option for making intermediate directories as required, so that you can create, e.g., **~/foo** and **~/foo/bar** with a single command? *Hint*: Refer to the man page for **mkdir**.

2. Use the option from the previous exercise to make the directory **foo** and, within it, the directory **bar** (i.e., **~/foo/bar**) with a single command.

3. By piping the output of **ls** to **grep**, list everything in the home directory that contains the letter "o".

4.3 Navigating Directories

We saw in Section 4.2 how to use **cd** to change to a directory with a given name. This is one of the most common ways of navigating, but there are a couple of special forms worth knowing. The first is changing to the directory one level up in the hierarchy using **cd ..** (read "see-dee dot-dot"):

```
$ pwd
/Users/mhartl/text_files
$ cd ..
$ pwd
/Users/mhartl
```

In this case, because **/Users/mhartl** is my *home directory*, we could have accomplished the same thing using **cd** by itself:

```
$ cd text_files/
$ pwd
/Users/mhartl/text_files
$ cd
$ pwd
/Users/mhartl
```

The reason this works is that **cd** by itself changes to the user's home directory, wherever that is. This means that

```
$ cd
```

and

```
$ cd ~
```

are equivalent.

When changing directories, it's frequently useful to be able to specify the home directory somehow. For example, suppose we make a second directory and **cd** into it:

```
$ pwd
/Users/mhartl
$ mkdir second_directory
$ cd second_directory/
```

Now if we want to change to the **text_files** directory, we can **cd** to **text_files** via the home directory **~**:

```
$ pwd
/Users/mhartl/second_directory
$ cd ~/text_files
$ pwd
/Users/mhartl/text_files
```

Incidentally, we're now in a position to understand the prompts shown in Figure 1.6: I have my prompt configured to show the current directory, which might be something like **[~]**, **[ruby]**, or **[projects]**. (We'll discuss how to customize the prompt in Part II (Section 6.7.1). Especially eager readers can exercise their technical sophistication (Box 1.4) by Googling for how to do it.)

Closely related to **..** for "one directory up" is **.** (read "dot") which means "the current directory". The most common use of **.** is when moving or copying files to the current directory:

```
$ pwd
/Users/mhartl/text_files
$ cd ~/second_directory
$ ls
$ cp ~/text_files/sonnets.txt .
$ ls
sonnets.txt
```

Note that the first call to **ls** returns nothing, because **second_directory** is initially empty.

Another common use of **.** is in conjunction with the **find** command, which like **grep** is incredibly powerful, but in my own use it looks like this 99% of the time:

```
$ cd
$ find . -name '*.txt'
./text_files/sonnet_1.txt
./text_files/sonnet_1_reversed.txt
./text_files/sonnets.txt
```

In words, what this does is find files whose names match the pattern ***.txt**, starting in the current directory **.** and then in its subdirectories.[2] The **find** utility is incredibly useful for finding a misplaced file at the command line.

Perhaps my favorite use of **.** is in "open dot", which will work only on macOS:

```
$ cd ~/ruby/projects
$ open .
```

The remarkable **open** command opens its argument using whatever the default program is for the given file or directory. (A similar command, **xdg-open**, works on some Linux systems.) For example, **open foo.pdf** would open the PDF file with the default viewer (which is Preview on most Macs). In the case of a directory such as **.**, that default program is the Finder, so **open .** produces a result like that shown in Figure 4.1.

A final navigational command, and one of my personal favorites, is **cd -**, which **cd**s to the *previous* directory, wherever it was:

```
$ pwd
/Users/mhartl/second_directory
$ cd ~/text_files
$ pwd
/Users/mhartl/text_files
$ cd -
/Users/mhartl/second_directory
```

I find that **cd -** is especially useful when combining commands, as described in Box 4.2.

---

2. My directory has a huge number of text files, 'cause that's just how I roll, so the command I ran was actually **find . -name '*.txt' | grep text_files**, which filters out anything that doesn't match the directory being used in this tutorial.

Box 4.2: Combining Commands

It's often convenient to combine commands at the command line, such as when installing software using the Unix programs `configure` and `make`, which often appear in the following sequence:

```
$ ./configure ; make ; make install
```

This line runs the `configure` program from the current directory `.`, and then runs both `make` and `make install`. (You are not expected to understand what these programs do, and indeed they won't work on your system unless you happen to be in the directory of a program designed to be installed this way.) Because they are separated by the semicolon character `;`, three commands are run in sequence.

An even better way to combine commands is with the double-ampersand `&&`:

```
$ ./configure && make && make install
```

The difference is that commands separated by `&&` run only if the previous command *succeeded*. In contrast, with `;` all the commands will be run no matter what, which will cause an error in the likely case that subsequent commands depend on the results of the ones that precede them.

I especially like to use `&&` in combination with `cd -`, which lets me do things like this:

```
$ build_article && cd ~/tau && deploy && cd -
```

Again, you are not expected to understand these commands, but the general idea is that we can (say) build an article in one directory, `cd` to a different directory, deploy (perhaps a website (https://tauday.com/)) to production, and then `cd` back (`cd -`) to the original directory, where we can continue our work. Then, if need be, we can just use up arrow (or one of the techniques from Box 3.1) to retrieve the whole thing and do it all again.

4.3.1 Exercises

1. How do the effects of **cd** and **cd ~** differ (or do they)?

2. Change to **text_files**, then change to **second_directory** using the "one directory up" double-dot operator `..`.

3. From wherever you are, create an empty file called **nil** in **text_files** using whatever method you wish.

4. Remove **nil** from the previous exercise using a different path from the one you used before. (In other words, if you used the path **~/text_files** before, use something like **../text_files** or **/Users/<username>/text_files**.)

4.4 Renaming, Copying, and Deleting Directories

The commands for renaming, copying, and deleting directories are similar to those for files (Section 2.3), but there are some subtle differences worth noting. The command with the least difference is **mv**, which works just as it does for files:

```
$ mkdir foo
$ mv foo/ bar/
$ cd foo/
-bash: cd: foo: No such file or directory
$ cd bar/
```

Here the error message indicates that the **mv** worked—there is no file or directory called **foo**:

```
$ cd foo/
-bash: cd: foo: No such file or directory
```

(The word **bash** refers to the name of the particular shell program being run, which in this case is the "Bourne Again SHell".) The only minor subtlety is that the trailing slashes (which are typically added automatically by tab completion (Box 2.4)) are optional:

```
$ cd
$ mv bar foo
$ cd foo/
```

This issue with trailing slashes never makes a difference with **mv**, but with **cp** it can be a source of much confusion. In particular, when copying directories, the behavior you usually want is to copy the directory contents *including* the directory, which on many systems requires leaving off the trailing slash. When copying files, you also need to include the **-r** option (for "recursive"). For example, to copy the contents of the

text_files directory to a new directory called **foobar**, we use the command shown in Listing 4.1.

Listing 4.1: Copying a directory.

```
$ cd
$ mkdir foobar
$ cd foobar/
$ cp -r ../text_files .
$ ls
text_files
```

Note that we've used **..** to make a *relative path*, going up one directory and then into **text_files**. Also note the *lack* of a trailing slash in **cp -r ../text_files .**; if we included it, we'd get Listing 4.2 instead.

Listing 4.2: Copying with a trailing slash.

```
$ cp -r ../text_files/ .
$ ls
sonnet_1.txt       sonnet_1_reversed.txt sonnets.txt    text_files
```

In other words, Listing 4.2 copies the individual files, but not the directory itself. As a result, I recommend always omitting the trailing slash, as in Listing 4.1; if you want to copy only the files, be explicit using the star operator, as in:

```
$ cp ../text_files/* .
```

Unlike renaming (moving) and copying directories, which use the same **mv** and **cp** commands used for files, in the case of removing directories there's a dedicated command called **rmdir**. In my experience, though, it rarely works, as seen here:

```
$ cd
$ rmdir second_directory
rmdir: second_directory/: Directory not empty
```

Figure 4.2: This superhero understands how to use the power of `rm -rf` responsibly.

The error message here is what happens 99% of the time when I try to remove directories, because **rmdir** requires the directory to be empty. You can of course empty it by hand (using **rm** repeatedly), but this is frequently inconvenient, and I almost always use the more powerful (but *much* more dangerous) "remove recursive force" command **rm -rf**, which removes a directory, its files, and any subdirectories without confirmation (Listing 4.3).

Listing 4.3: Using `rm -rf` to remove a directory.

```
$ rm -rf second_directory/
$ ls second_directory
ls: second_directory: No such file or directory
```

As the error message from **ls** in Listing 4.3 indicates ("No such file or directory"), our use of **rm -rf** succeeded in removing the directory.

The powerful command **rm -rf** is too convenient to ignore, but remember: "With great power comes great responsibility" (Figure 4.2).[3]

---

3. Image courtesy of MeskPhotography/Shutterstock.

4.4.1 Grep Redux

Now that we know a little about directories, we are in a position to add a useful **grep** variation to our toolkit from Section 3.4. As with **cp** and **rm**, **grep** takes a "recursive" option, **-r**, which in this case greps through a directory's files and the files in its subdirectories. This is incredibly useful when you're looking for a string in a file somewhere in a hierarchy of directories, but you're not sure where the file is. Here's the setup, which puts the word "sesquipedalian" in a file called **long_word.txt**:

```
$ cd text_files/
$ mkdir foo
$ cd foo/
$ echo sesquipedalian > long_word.txt
$ cd
```

The final **cd** puts us back in the home directory. Suppose we now want to find the file containing "sesquipedalian". The way *not* to do it is this:

```
$ grep sesquipedalian text_files      # This doesn't work.
grep: text_files: Is a directory
```

Here **grep**'s error message indicates that the command didn't work, but adding **-r** does the trick:

```
$ grep -r sesquipedalian text_files
text_files/foo/long_word.txt:sesquipedalian
```

Because we don't usually care about case when searching files, I recommend making a habit of adding the **-i** option when grepping recursively, as follows:

```
$ grep -ri sesquipedalian text_files
text_files/foo/long_word.txt:sesquipedalian
```

Armed with **grep -ri**, we are now equipped to find strings of our choice in arbitrarily deep hierarchies of directories.

4.4.2 Exercises

1. Make a directory **foo** with a subdirectory **bar**, then rename the subdirectory to **baz**.

2. Copy all the files in **text_files**, *with* directory, into **foo**.

3. Copy all the files in **text_files**, *without* directory, into **bar**.

4. Remove **foo** and everything in it using a single command.

4.5 Summary

Important commands from this chapter are summarized in Table 4.1.

Table 4.1: Important commands from Chapter 4.

Command	Description	Example
mkdir <name>	Make directory with name	`$ mkdir foo`
pwd	Print working directory	`$ pwd`
cd <dir>	Change to <dir>	`$ cd foo/`
cd ~/<dir>	cd relative to home	`$ cd ~/foo/`
cd	Change to home directory	`$ cd`
cd -	Change to previous directory	`$ cd && pwd && cd -`
.	The current directory	`$ cp ~/foo.txt .`
..	One directory up	`$ cd ..`
find	Find files & directories	`$ find . -name foo*.*`
cp -r <old> <new>	Copy recursively	`$ cp -r ~/foo .`
rmdir <dir>	Remove (empty) dir	`$ rmdir foo/`
rm -rf <dir>	Remove dir & contents	`$ rm -rf foo/`
grep -ri <string> <dir>	Grep recursively (case-insensitive)	`$ grep -ri foo bar/`

4.5.1 Exercises

1. Starting in your home directory, execute a single command-line command to make a directory **foo**, change into it, create a file **bar** with content "baz", print out **bar**'s contents, and then **cd** back to the directory you came from. *Hint*: Combine the commands as described in Box 4.2.

2. What happens when you run the previous command again? How many of the commands executed? Why?

3. Explain why the command `rm -rf /` is unbelievably dangerous, and why you should never type it into a terminal window, not even as a joke.

4. How can the previous command be made even more dangerous? *Hint*: Refer to Box 4.1. (This command is so dangerous you shouldn't even *think* it, much less type it.)

4.6 Conclusion

Congratulations! You've officially learned enough command line to be *dangerous*. Of course, this is only one step on a longer journey, both toward command-line excellence (Box 4.3) and software development wizardry. As you proceed on this journey, you will probably discover that learning computer magic can be exciting and empowering, but it can also be *hard*. Indeed, you may already have discovered this fact, either on your own or while following Part I of *Learn Enough Developer Tools to Be Dangerous*. To those brave magicians-in-training who wish to proceed, I offer the following sequence:

1. *Learn Enough Developer Tools to Be Dangerous*
   (a) Part I: *Learn Enough Command Line to Be Dangerous* (https://www .learnenough.com/command-line) (you are here)
   (b) Part II: *Learn Enough Text Editor to Be Dangerous* (https://www.learnenough .com/text-editor)
   (c) Part III: *Learn Enough Git to Be Dangerous* (https://www.learnenough .com/git)

2. **Web Basics**
   (a) *Learn Enough HTML to Be Dangerous* (https://www.learnenough.com /html)
   (b) *Learn Enough CSS & Layout to Be Dangerous* (https://www.learnenough.com /css-and-layout)
   (c) *Learn Enough JavaScript to Be Dangerous* (https://www.learnenough.com /javascript)

3. **Application Development**

    (a) *Learn Enough Ruby to Be Dangerous* (https://www.learnenough.com/ruby)

    (b) *Ruby on Rails Tutorial* (https://www.railstutorial.org/)

    (c) *Learn Enough Action Cable to Be Dangerous* (https://www.learnenough.com/action-cable) (optional)

Box 4.3: Additional Resources

I recommend following the Learn Enough sequence described in the main text, as it represents the shortest path to technical proficiency and software development skills, but at some point you'll probably want to expand your command-line skills as well. When that time comes, I suggest consulting this list, which consists mainly of resources found to be valuable by readers of the present tutorial:

- *Conquering the Command Line* book and screencasts by Mark Bates (https://conqueringthecommandline.com/)
- edX course on Linux (https://www.edx.org/course/introduction-to-linux)
- Codecademy course on the command line (https://www.codecademy.com/learn/learn-the-command-line)
- *Learning the Shell* (http://linuxcommand.org/lc3_learning_the_shell.php)

PART II
Text Editor

CHAPTER 5

Introduction to Text Editors

Learn Enough Text Editor to Be Dangerous (https://www.learnenough.com/text
-editor)—Part II of *Learn Enough Developer Tools to Be Dangerous*—is designed to help
you learn to use what is arguably the most important item in the aspiring computer
magician's bag of tricks (Box 1.1): a *text editor* (Figure 5.1). Learning how to use a text
editor is an essential component of technical sophistication.

Unlike other text editor tutorials, which are typically tied to a specific editor,
Part II of *Learn Enough Developer Tools to Be Dangerous* is designed to introduce the
entire *category* of application—a category many people don't even know exists.[1] Moreover, editor-specific tutorials tend to be aimed at professional developers, and generally
assume years of experience, but *Learn Enough Text Editor to Be Dangerous* doesn't even
assume you know what a "text editor" is. Its only prerequisite is a basic understanding
of the Unix command line, such as that provided by Part I.[2]

Unlike most programs used to produce written documents, such as word processors and email clients, a *text editor* is an application specifically designed to edit
plain text (often called just "text" for short). Learning to use a text editor is important because text is ubiquitous in modern computing—it's used for code, markup,
configuration files, and many other things.[3] (Indeed, I'm using plain text to write this
very document.) Although it's surprisingly difficult to define exactly what "plain text"

---

1. This is why we refer to this part of *Learn Enough Developer Tools to Be Dangerous* as "Learn Enough Text
Editor" rather than "Learn Enough Text Editors"—saying "text editor" in the singular emphasizes that we
are discussing the general category and not just a series of specific editors.

2. This is required both because we'll be launching text editors from the command line and because some
of the examples involve customizing and extending the *shell program* in which the command line runs.

3. For more on the power of text, see the insightful post "always bet on text" (https://graydon2.dreamwidth
.org/193447.html?HN2).

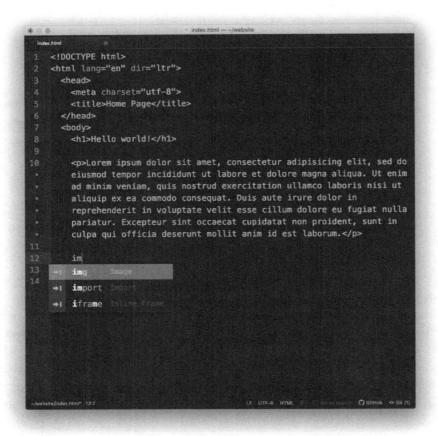

Figure 5.1: A text editor.

is, from a practical perspective it means that the text itself doesn't have any formatting, or at least none that matters. There's no notion of *emphasized* or **boldface** text, the font size and **typeface** don't matter, etc.—the only thing that does matter is the *content*. For example, although the previous sentence contains formatted output like *this*, its source is plain text, and appears as in Listing 5.1.[4]

---

4. Technically, the em dash "—" appears as a raw Unicode character rather than as —, but the latter is equivalent and is easier to notice in a code listing. For similar reasons, Listing 5.1 uses regular quotes in place of fancier "curly" quotes. (On most browsers, setting Unicode to display properly requires a full HTML document with the proper headers. These sorts of considerations are covered in *Learn Enough HTML to Be Dangerous* (https://www.learnenough.com/html).)

Listing 5.1: The HTML source of a sentence in this tutorial.

```
There's no notion of <em>emphasized</em> or <strong>boldface</strong> text,
the <small>font size</small> and <code>typeface</code> don't matter,
etc.&mdash;the only thing that does matter is the <em>content</em>.
```

In Listing 5.1, the desired formatting options are indicated with special *tags* (such as the HTML emphasis tag **...**) rather than by changing the appearance of the text itself.[5] This is the main reason why the more familiar word processor programs such as Word aren't well-suited to editing plain text, and a different sort of tool is needed (Box 5.1).

Box 5.1: Word Processors vs. Text Editors

Even if you've never used a text editor, the chances are good that you've used a similar tool, a *word processor*. There's a lot of overlap between the features of word processors and text editors. For example, they both allow you to create documents, find and replace or cut/copy/paste text, and save the results. The main difference is that word processors are generally designed to produce documents following the principle of "What You See Is What You Get" (WYSIWYG, pronounced "WIZ-ee-wig"), so that effects such as *emphasis* or **boldface** are achieved directly in the application, instead of with plain-text markup like emphasis or **boldface**. For the most part, word processors also save their results in proprietary formats that sometimes go bad (as many who've tried opening old Word files have learned to their chagrin).

In contrast, text editors are designed to modify plain text, one of the most universal and durable formats. Text editors also differ from word processors in having features aimed at more technical users, including syntax highlighting for source code (Section 7.2.1), automatic indentation (Section 7.2.3), support for regular expressions (Section 7.4.3), and customization via packages and snippets (Section 7.5). A good text editor is thus an essential tool in every technical person's toolkit.

---

5. It is up to the individual application to determine how to display the formatting. For example, HTML is designed to be rendered and displayed by web browsers like Chrome and Safari, which typically display emphasized text using *italics*.

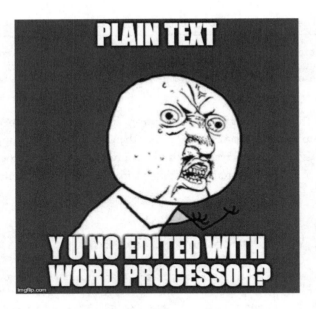

Figure 5.2: Why not edit plain text with Microsoft Word?

Building on the material developed in Part I, Part II starts by covering the important *Vim* editor (Section 5.1), which can be run at the command line directly inside a terminal window. Vim will give us a chance to see our first examples of the most important functions of a text editor, but because Vim can be forbiddingly complex for a beginner, in this tutorial we will cover only the bare minimum necessary to make basic edits. The rest of the tutorial will expand on the themes developed in this chapter by describing some of the many powerful features required in any programmer-grade text editor, with examples drawn principally from *Atom*, an open-source cross-platform editor, with concepts applicable to the closely related Sublime Text and Visual Studio Code editors and to Cloud9, a cloud IDE.[6]

As with *Learn Enough Command Line to Be Dangerous* (https://www.learnenough .com/command-line), this tutorial is part of the *Unix tradition*, which includes virtually every operating system you've ever heard of (macOS, iOS, Android, Linux, etc.) except Microsoft Windows. Although all the editors we'll discuss do run under

---

6. Some developers use an *integrated development environment*, or IDE, for their day-to-day programming, but every IDE includes an integrated text editor, so the lessons in this tutorial still apply.

Windows, using a non-Unix OS introduces friction into the process, so Microsoft Windows users are encouraged to set up a Linux-compatible development environment by following the steps in Section A.3.3 or by using the Linux-based cloud IDE discussed in Section A.2.

The focus throughout *Learn Enough Text Editor to Be Dangerous* is on general principles, so no matter which editor you end up using, you will have a good mental checklist of the kinds of tasks you should rely on your editor to perform. In addition, because the details vary by particular text editor and by system, this tutorial presents an ideal opportunity to continue developing your *technical sophistication* (Box 5.2). Finally, don't feel any pressure to master everything at once. You can be productive with even a small subset of what's included in this tutorial. Because technically sophisticated people use text editors practically every day, you'll keep learning new tricks in perpetuity.

Box 5.2: Technical Sophistication

The phrase *technical sophistication*, mentioned before in Part I (Box 1.4), refers to the general ability to use computers and other technical things. This includes both existing knowledge (such as familiarity with text editors and the Unix command line) and the ability to acquire *new* knowledge, as illustrated in "Tech Support Cheat Sheet" (https://m.xkcd.com/627/) from xkcd. Unlike "hard skills" like coding and version control, this latter aspect of technical sophistication is a "soft skill"—difficult to teach directly, but essential to develop if you want to work with computer programmers or to become a programmer yourself.

In the context of text editors, technical sophistication includes things like reading menu items to figure out what they do, using the Help menu to discover new commands, learning keyboard shortcuts by reading menu items or Googling around, etc. It also involves a tolerance for ambiguity: Technically sophisticated readers won't panic if a tutorial says to use ⌘Z to Undo something when it's actually ^Z on their system. They also won't panic if they see ⌘Z but don't know what ⌘ means, because they know they can skim ahead to find something like Table 1.1 (or simply Google for "mac special keys"). Perhaps the most important aspect of technical sophistication is an *attitude*—a confidence and can-do spirit in the face of confusion that is well worth cultivating.

Throughout the rest of *Learn Enough Text Editor to Be Dangerous*, we'll refer back to this box whenever we encounter examples of issues that require a little technical sophistication to solve. With experience, you too will become one of the "computer people" from "Tech Support Cheat Sheet" who seem to have the magical ability (Box 1.1) to figure out technical things. (*Warning*: You might need a new shirt (Figure 5.3).)

5.1 Minimum Viable Vim

The vi (pronounced "vee-eye") editor dates back to the earliest days of Unix, and Vim (pronounced "vim") is an updated version that stands for "Vi IMproved". Vim is absolutely a full-strength text editor, and many developers use it for their daily editing needs, but the barrier to Vim mastery is high, and it requires substantial customization and technical sophistication (Box 5.2) to reach its full potential. Vim also has a large and often obscure set of commands, which rarely correspond to native keybindings

Figure 5.3: A T-shirt for the technically sophisticated.

(keyboard shortcuts), making Vim challenging to learn and remember. As a result, I generally recommend beginners learn a "modern" editor (Chapter 6) for everyday use. Nevertheless, I consider a minimal proficiency with Vim to be essential, simply because Vim is utterly ubiquitous—it's present on virtually every Unix-like system in the known universe, which means that if you ssh into some random server halfway 'round the globe, Vim will probably be there.

This chapter includes only Minimum Viable Vim—just enough to use Vim to do things like edit small configuration files or Git commits.[7] It's not even really enough to be *dangerous*. But it's worth noting that even mastering Minimum Viable Vim puts you in elite company—because Vim is so difficult, even a little Vim knowledge is the sort of thing that can impress your friends (or a job interviewer).

Note: If you're using macOS, you should follow the instructions in Box 2.3 at this time.

5.2 Starting Vim

Unlike most of the modern editors discussed starting in Chapter 6, Vim can be run directly inside a terminal window, and requires no graphical interface.[8] All you do is type **vim** at the prompt:

```
$ vim
```

Typical results of running the **vim** command appear in Listing 5.2 and Figure 5.4. In both cases, the tildes (~) are not characters in the file but rather represent lines that have yet to be defined.

Listing 5.2: A textual representation of a Vim window (message and versions may differ).

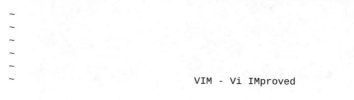

```
~
~
~
~
~
~                          VIM - Vi IMproved
```

---

7. See Part III for more details.

8. Vim thus dovetails nicely with a command-line tutorial like *Learn Enough Command Line to Be Dangerous* (Part I).

```
~
~
~                          version 7.3
~                     by Bram Moolenaar et al.
~              Vim is open source and freely distributable
~
~                      Help poor children in Uganda!
~          type  :help iccf<Enter>       for information
~
~          type  :q<Enter>               to exit
~          type  :help<Enter>  or  <F1>  for on-line help
~          type  :help version7<Enter>   for version info
```

If starting Vim is easy, learning to use it, at least at first, can be incredibly hard. This is mostly due to Vim being a *modal* editor, which is probably unlike anything you have used before (Box 5.3). Vim has two principal *modes*, known as *normal mode* and

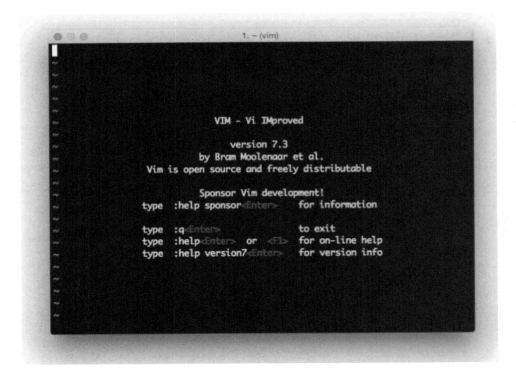

Figure 5.4: Vim running in a terminal window.

insertion mode. Normal mode is for doing things like moving around the file, deleting content, or finding and replacing text, whereas insertion mode is for inserting text.

Box 5.3: Modal Vim

When I first started to learn programming in the Unix tradition (as opposed to my childhood experience with Microsoft DOS, BASIC, and Pascal), I distinctly recall being absolutely mortified at the unbelievably primitive editor I was expected to use. At the time, I was a first-year undergraduate at Harvard University, working in a research group at the Harvard-Smithsonian Center for Astrophysics. The tool I had been handed was vi. To say that it seemed like a downgrade from word processors is a gross understatement (Figure 5.2).

What shocked me most about vi was modal editing: Unlike word processors, vi didn't let me just click in the window and start typing. Instead, there were a profusion of options (i, a, and o among them) for switching to *insertion mode*, and all it took was a few wrong keystrokes for all hell to break loose. Although the intervening years have seen a proliferation of more modern text editors, whose design is much more like the click-and-type interface I expected from my experience with word processors, the enduring popularity of vi (via Vim) means that learning the basics of modal editing is a valuable skill, even if it might at first seem ridiculously foreign.

Going back and forth between these two modes can cause a lot of confusion, especially since virtually all other programs that edit text (including word processors, email clients, and most text editors) have only insertion mode. Part of what makes Vim particularly confusing is that it *starts* in normal mode, which means that, if you try entering text immediately after starting Vim (as in Listing 5.2), the result will be chaos.

Because confusion is the likeliest result if you're not used to Vim's modal editing, we're going to start our study of Vim with the *Most Important Vim Command*™. One of my college friends, who was a huge partisan of vi's (and hence Vim's) historical rival Emacs (Box 5.4), claimed the Most Important Vim Command™ was the only one he ever wanted to learn. Here it is:

```
ESC:q!<return>
```

This command means "Press the Escape key, then type 'colon q exclamation–point', then press the return key." We'll learn in a moment what this does and why, but for now we'll start by practicing it a couple of times in the exercises.

Box 5.4: Holy Wars: vi vs. Emacs

The Jargon File (http://www.catb.org/jargon/html/) defines *holy wars* as follows:

holy wars: n.
[from *Usenet*, but may predate it; common] *flame wars* over *religious issues*. The paper by Danny Cohen that popularized the terms *big-endian* and *little-endian* in connection with the LSB-first/MSB-first controversy was entitled *On Holy Wars and a Plea for Peace*.

Great holy wars of the past have included *ITS* vs. *Unix*, *Unix* vs. *VMS*, *BSD* Unix vs. System V, *C* vs. *Pascal*, *C* vs. FORTRAN, etc. In the year 2003, popular favorites of the day are KDE vs. GNOME, vim vs. elvis, Linux vs. [Free|Net|Open]BSD. Hardy perennials include *EMACS* vs. *vi*, my personal computer vs. everyone else's personal computer, ad nauseam. The characteristic that distinguishes holy wars from normal technical disputes is that in a holy war most of the participants spend their time trying to pass off personal value choices and cultural attachments as objective technical evaluations. This happens precisely because in a true holy war, the actual substantive differences between the sides are relatively minor. See also *theology*.

As noted in the Jargon File entry, one of the longest-raging holy wars is fought between proponents of vi and its arch-rival Emacs (sometimes written "EMACS"), both of which have played important roles in the Unix computing tradition. Both also retain much popular support, although my guess is that, with the popularity of Vim, vi has taken a decisive lead in recent years. Of course, this is just the sort of statement that serves to perpetuate a holy war—likely prompting Emacs partisans to, say, make claims about the superior power and customizability of their favorite editor.

If you wanted to start a *new* holy war, you might try something like, "Happily, the vi vs. Emacs holy war is now mostly a historical curiosity, as anyone who's anyone has switched to a modern editor like Atom or Sublime." It's going to be quite a show—better bring some popcorn (Figure 5.5).[9]

---

9. Image courtesy of Cartno Studio/Shutterstock.

Figure 5.5: Watching a holy war play out can be entertaining.

5.2.1 Exercises

1. Start Vim in a terminal, then run the Most Important Vim Command™.

2. Restart Vim in a terminal. Before typing anything else, type the string "This is a Vim document." What happened? Confusing, right?

3. Use the Most Important Vim Command™ to recover from the previous exercise and return to the normal command-line prompt.

5.3 Editing Small Files

Now that we know the Most Important Vim Command™, we'll start learning how to use Vim for real by opening and editing a small file. In Section 5.2, we started by running **vim** by itself, but it's more common to use a filename as an argument. Let's navigate to the home directory of our system and then run such a command, which will either open the corresponding file (if it exists) or create it (if it doesn't):[10]

```
$ cd
$ vim .bashrc
```

Here **.bashrc** is a standard configuration file for Bash.[11]

As noted above, the **vim .bashrc** command will automatically create the corresponding **.bashrc** file if it doesn't already exist on your system. This important file is used to configure the *shell*, which is the program that supplies a command line—in this case, *Bash*, which is a pseudo-acronym that stands for *Bourne Again SHell* (also written as "Bourne-again shell").[12] (Recall from Box 2.3 that the default on macOS is now Zsh, so if you're on a Mac you should follow the instructions there to switch to Bash if you haven't already.)

As is common on Unix-based systems, the configuration file for Bash begins with a dot, indicating (as noted in Section 2.2) that the file is *hidden*. That is, it doesn't show up by default when listing directory contents with **ls**, or even when viewing the directory using a graphical file browser.

We'll learn in Section 5.4 how to save changes to this file, but for now we're just going to add some dummy content so that we can practice moving around. In Section 5.2, we learned that Vim starts in normal mode, which means that we can change location, delete text, etc. Let's go into insertion mode to add some content. The first step is to press the **i** key to *insert* text. Then, type a few lines (separated by returns), as shown in Listing 5.3.[13] (There may be other existing content, which you should simply ignore.)

---

10. Technically, the file isn't actually created until you save it (Section 5.4), but you get the idea.

11. See this Stack Overflow thread (https://stackoverflow.com/questions/902946/about-bash-profile -bashrc-and-where-should-alias-be-written-in) if you're curious about where the "rc" comes from.

12. The first program in the sequence was the Bourne shell; in line with the Unix tradition of terrible puns, its successor is called the "Bourne again" (as in "born again") shell.

13. Recall from Section 4.1 that a tilde ~ is used to indicate the home directory, so **~/.bashrc** in the Listing 5.3 caption refers to the Bash configuration file in the home directory.

Listing 5.3: Adding some text after typing **i** to insert.
~/.bashrc

```
1   lorem ipsum
2   dolor sit amet
3   foo bar baz
4   I've made this longer than usual because I haven't had time to make it shorter.
```

After entering the text in Listing 5.3, press **ESC** (the escape key) to switch from insertion mode back to normal mode.

Now that we have some text on a few lines, we can learn some commands for moving around small files. (We'll cover some commands for navigating large files in Section 5.6.) The easiest way to move around is to use arrow keys—up, down, left, right—which is what I recommend.[14] Vim has literally jillions of ways of moving around, and if you decide to use Vim as your primary text editor I recommend learning them, but for our purposes the arrow keys are fine. The only two additional commands I feel are essential are the ones to move to the beginning and end of the line, which are **0** and **$**, respectively.[15]

5.3.1 Exercises

1. Use the arrow keys to navigate to line 4 in the file from Listing 5.3.

2. Use the arrow keys to go to the end and then the beginning of line 4. Cumbersome, eh?

3. Go to the beginning of line 4 by using the command mentioned in the text.

4. Go to the end of line 4 using the command mentioned in the text.

5.4 Saving and Quitting Files

Having learned a little about moving around and inserting text, now we're going to learn how to save a file. Our specific example will involve making a useful new Bash

---

14. Vim purists will tell you that there's a better way, namely, to use the **h**, **j**, **k**, **l** to move around, but it takes a lot of practice for this to become intuitive, and it's certainly not necessary to be *dangerous*.

15. These are not native keybindings (on macOS they would be Command–left arrow and Command–right arrow), which as noted in the introduction to Chapter 5 makes it harder to learn them. This is just one of many reasons I don't generally recommend beginners use Vim as their primary editor.

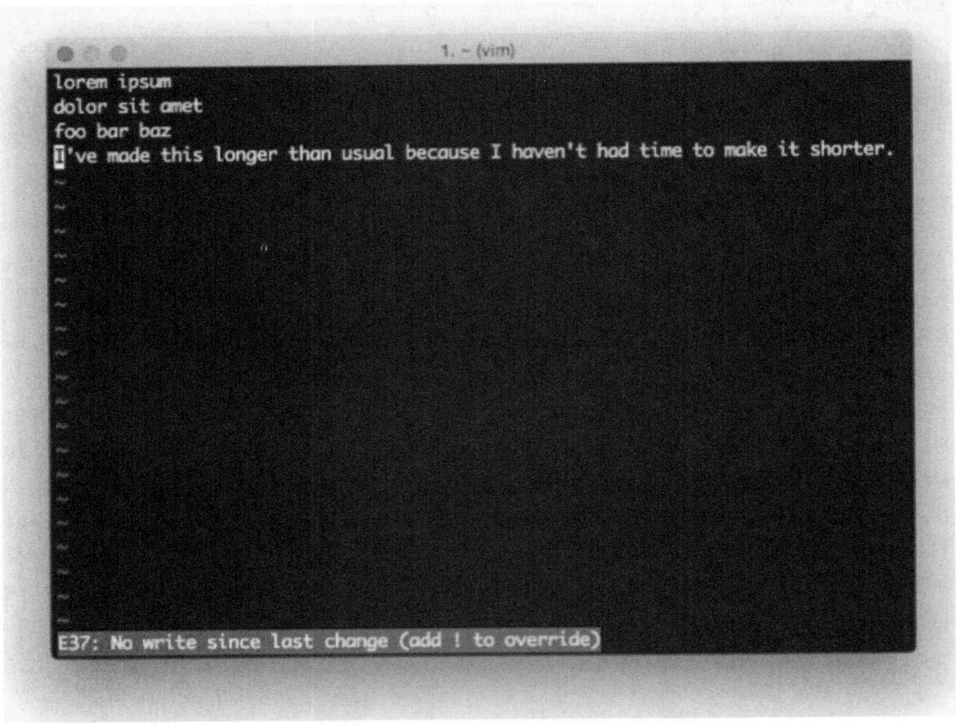

Figure 5.6: Trying to quit a file with unsaved changes.

command, but first we have to deal with the current state of the Bash profile file. The text we added in Listing 5.3 is gibberish (at least from Bash's perspective), so what we'd like to do is quit the file without saving any changes. For historical reasons, some Vim commands (especially those having to do with file manipulation) start with a colon **:**, and the normal way to quit a file is with **:q<return>**, but that only works when there are no changes to save. In the present case, we get the error message "No write since last change (add ! to override)", as shown in Figure 5.6.

Following the message's advice, we can type **:q!<return>** to force Vim to quit without saving any changes (Figure 5.7), which returns us to the command line.

You may have noticed that we're now in a position to understand the Most Important Vim Command™ introduced in Section 5.2: No matter what terrible things you might have done to a file, as long as you type **ESC** (to get out of insertion mode if

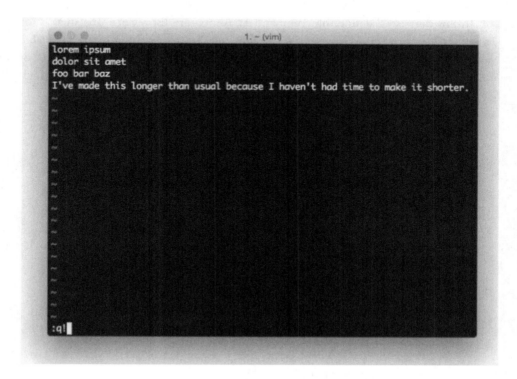

Figure 5.7: Forcing Vim to quit.

necessary)[16] followed by `:q!<return>` (to force-quit) you are guaranteed not to do any damage.

Of course, Vim is only really useful if we can save our edits, so let's add some useful text and then write out the result. As in Section 5.3, we'll work on the `.bashrc` file, and the edit we'll make will add an *alias* to our shell. In a computing context, an alias is simply a synonym for a command or set of commands. The main use for Bash aliases is defining shorter commands for commonly used combinations.[17]

---

16. Hitting **ESC** while in normal mode does no harm, so it's a good idea to include this step in any case.

17. To learn how to write aliases using Zsh, see "Using Z Shell on Macs with the Learn Enough Tutorials" (https://news.learnenough.com/macos-bash-zshell). TL;DR: The syntax is identical; the only difference is that you edit a file called `.zshrc` instead of `.bashrc`. (Indeed, in Bash you can actually edit a file called `.bashrc` instead, which makes the parallel even clearer.)

In this case, we'll define the command **lr** (short for "list reverse") as an alias for **ls -hartl**, which is the command to list files and directories using **h**uman-readable values for the sizes (e.g., 29K instead of 29592 for a 29-kilobyte file), including **a**ll of them (even hidden ones), ordered by **r**everse **t**ime, **l**ong form. This command, which as you may recognize from an exercise in Section 3.1.1, is useful for seeing which files and directories have recently changed (as well as being, for obvious reasons, one of my personal favorites). After defining the alias, we'll be able to replace the more verbose

```
$ ls -hartl
```

with the pithier

```
$ lr
```

The steps appear as follows:

1. Press **i** to enter insertion mode.
2. Enter the contents shown in Listing 5.4. (On some systems, the **.bashrc** file may include some pre-existing content, which you can simply leave in place.)
3. Press **ESC** to exit insertion mode.
4. Write the file using **:w<return>**.
5. Quit Vim by typing **:q<return>**.

Note: If you make any mistakes, you can type **ESC** followed by **u** to *undo* any of the previous steps. (Most programs use Command-Z or Ctrl-Z to undo things, yet another example of the non-native keybindings used by Vim. In contrast, the editors discussed starting in Chapter 6 all support native keybindings.)

Listing 5.4: Defining a Bash alias.
~/.bashrc

```
alias lr='ls -hartl'
```

After adding the **lr** alias to **.bashrc**, writing the file, and quitting, you may be surprised to find that the command doesn't yet work:

```
$ lr
-bash: lr: command not found
```

This is because we need to tell the shell about the updated Bash profile file by
"sourcing" it using the **source** command, as shown in Listing 5.5.[18]

Listing 5.5: Activating the alias by sourcing the Bash profile.

```
$ source .bashrc
```

With that, the **lr** command should work as advertised:

```
$ lr
.
.
.
drwx------+  15 mhartl   staff    510B Sep  4 18:58 Desktop
-rw-------    1 mhartl   staff     13K Sep  4 19:13 .viminfo
-rw-r--r--    1 mhartl   staff     46B Sep  4 19:14 .bashrc
drwxr-xr-x+ 117 mhartl   staff    3.9K Sep  4 19:14 .
```

By the way, the **.bashrc** file is sourced automatically when we open a new terminal
tab or window, so explicit sourcing is necessary only when we want a change to be
reflected in the *current* terminal.

5.4.1 Exercises

1. Define an alias **g** for the commonly used *case-insensitive grep* **grep -i**. What happens if, after making your changes and hitting **ESC**, you issue the command **:wq**
   instead of **:w** and **:q** separately?

2. You may recall the **curl** command from Section 3.1, which lets us interact with
   URLs via the command line. Define **get** as an alias for **curl -OL**, which is
   the command to download a file to the local disk (while following any redirects
   encountered along the way).

---

18. Incidentally, a bare "dot" is a shorthand for **source**, so in fact you can type **. .bashrc** to obtain the
same result. (This usage is unrelated to the use of a dot to refer to the current directory (Section 4.3).)

3. Use the alias from the previous exercise to execute the command shown in Listing 5.6, which downloads a longer text file for use in Section 5.6.

Listing 5.6: Downloading a longer text file for use in a future section.

```
$ get cdn.learnenough.com/sonnets.txt
```

5.5 Deleting Content

As with every category of text manipulation, Vim has an enormous number of commands for deleting content, but in this section we're just going to cover the absolute minimum. We'll start with deleting single characters, which we can do in normal mode using the **x** command:

1. Open **.bashrc** and insert the misspelled word **aliaes**.
2. Get back to normal mode by pressing **ESC**.
3. Move the cursor over the **e** in **aliaes** (Figure 5.8) and press **x**.

There are lots of fancy ways to delete text, but by repeatedly pressing **x** it's easy (if a bit cumbersome) to delete entire words or even entire lines. On the other hand, deleting lines is enough of a special case to merit inclusion. Let's get rid of the extra **alias** we added by pressing **dd** to delete the line. Voilà! It should be gone (Figure 5.9). To get it back, you can press **p** to "put" the line, which also allows you to simulate copying and pasting one line at a time. (Again, this is a minimal subset of Vim; if you decide to get good at it, you'll learn lots of better ways to do things.)

5.5.1 Exercises

1. Using Vim, open a new file called **foo.txt**.
2. Insert the string "A leopard can't change it's spots." (Figure 5.10[19])
3. Using the **x** key, delete the character necessary to correct the mistake in the line you just entered. (If you can't find the error, refer to Table 5.1.)

---

19. Image courtesy of apoplexia/123rf.com.

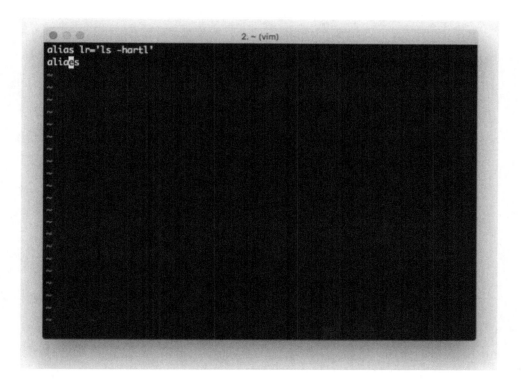

Figure 5.8: Preparing to delete a letter using **x**.

4. Use **dd** to delete the line, then use **p** to paste it repeatedly into the document.

5. Save the document and quit using a single command. *Hint*: See the first exercise in Section 5.4.1.

5.6   Editing Large Files

The final skills needed for your Minimum Viable Vim involve navigating large files. If you didn't download **sonnets.txt** in the exercises from Section 5.4, you should so do now (Listing 5.7).[20]

_____

20. If you completed the exercises in Section 6.7.1, you can use your own custom **get** alias in place of **curl -OL**.

Figure 5.9: The result of deleting a line with **dd**.

Listing 5.7: Downloading Shakespeare's *Sonnets*.

```
$ curl -OL https://cdn.learnenough.com/sonnets.txt
```

The resulting file contains the full text of Shakespeare's *Sonnets*, which is 2620 lines, 17670 words, and 95635 characters long, which we can verify using the word count command **wc** (Section 3.2.1):

```
$ wc sonnets.txt
   2620   17670   95635 sonnets.txt
```

Figure 5.10: This animal's spot-changing abilities are frequently questioned.

On many systems, Vim shows some of the same basic stats upon opening the file:

```
$ vim sonnets.txt
```

The result on my system appears in Figure 5.11. Because of its length, this file is far too long to navigate conveniently by hand.

As before, there are lots of commands for moving around, but I find that the most useful ones involve moving a screen at a time, moving to the beginning or end, or searching. The commands to move one screen at a time are Ctrl-F (Forward) and Ctrl–B (Backward). To move to the end of the file, we can use **G**, and to move to the beginning we can use **1G**. Finally, perhaps the most powerful navigation command is *search*, which involves typing slash **/** followed by the string you want to find. The trick

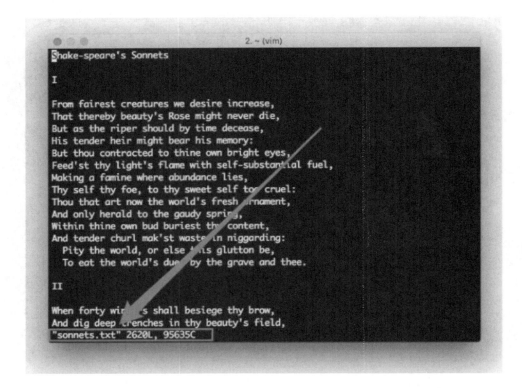

Figure 5.11: Some file stats displayed upon starting Vim.

is to type **/<string>** followed by return, and then press **n** to go to the next match
(if any).

This might all sound a little familiar, because it's the same as the interface to the
less program (Section 3.3).[21] This is one of the advantages of learning basic Unix
commands: Many of the patterns recur in many different contexts.

---

21. Depending on your system, there may be minor differences between the Vim and **less** interfaces. For
example, on my system the slash operator is case-sensitive when used with **less** and case-insensitive when
used with Vim. As usual, use your technical sophistication (Box 5.2) to resolve any discrepancies.

5.6.1 Exercises

1. With **sonnets.txt** open in Vim, move down three screens and then back up three screens.

2. Go to the end of the file. What is the last line of the final sonnet?

3. Navigate back to the top to change the old-style name "Shake-speare" on line 1 of **sonnets.txt** to the more modern "Shakespeare", and save the result.

4. Use Vim's search feature to discover which sonnet contains references to Cupid, the Roman god of love.

5. Confirm that **18G** goes to the final line of the first sonnet. What do you suppose that command does? *Hint*: Recall that **1G** goes to the beginning of the file, i.e., line 1.

5.7 Summary

Important commands from this chapter are summarized in Table 5.1. If you're interested in learning more about Vim, dropping "learn vim" into a search engine is a good bet. The Interactive Vim tutorial (https://www.openvim.com/) is especially recommended.

5.7.1 Exercises

1. Open **sonnets.txt**.

2. Go to the last line.

3. Go to the end of the last line.

4. Make a new line that says "That's all, folks! Bard out. <drops mic>". Make sure to move the cursor one space to the right so you don't drag the final period along.

5. Write out the file.

6. Undo your changes.

7. Write out and quit the file.

8. Reopen the file and type **2620dd**.

9. Realize that you just deleted the entire file contents, and apply the Most Important Vim Command™ to ensure that no damage is done.

Table 5.1: Important Vim commands from Chapter 5.

Command	Description
`ESC:q!<return>`	The Most Important Vim Command™
`i`	Exit normal mode, enter insertion mode
`ESC`	Exit insertion mode, enter normal mode
Arrow keys	Move around
`0`	Go to beginning of line
`$`	Go to end of line
`:w<return>`	Save (write) a file
`:q<return>`	Quit a file (must be saved)
`:wq<return>`	Write and quit a file
`:q!<return>`	Force-quit a file, discarding any changes
`u`	Undo
`x`	Delete the character under the cursor
`dd`	Delete a line
`p`	Put (paste) deleted text
`it's spots`	No, you mean `its spots`
`Ctrl-F`	Go forward one screen
`Ctrl-B`	Go backward one screen
`G`	Go to last line
`1G`	Go to first line
`/<string>`	Search for `<string>`

CHAPTER 6
Modern Text Editors

Having learned the minimal basics of text editing with Vim, we're now in a good position to appreciate the preferred "modern" text editors mentioned in Section 5.1. These editors include cross-platform native editors such as Sublime Text (https://www.sublimetext.com/), Visual Studio Code (VSCode) (https://code .visualstudio.com/), and Atom (https://atom.io/), and editors in the cloud like Cloud9 (https://aws.amazon.com/cloud9/). Modern editors are distinguished by their combination of power and ease of use: You can do fancy things like global search-and-replace, but (unlike Vim) they let you just click in a window and start typing. In addition, many of them (including Atom and Sublime) include an option to run in Vim compatibility mode, so even if you end up loving Vim you can still use a modern editor without having to give Vim up entirely.

Throughout the rest of this tutorial, we'll explore the capabilities of modern text editors. We'll end up covering all of the topics encountered in our discussion of Vim (Chapter 5), as well as many more advanced subjects, including opening files, moving around, selecting text, cut/copy/paste, deleting and undoing, saving, and finding/replacing—all of which are important for day-to-day editing. We'll also discuss both menu items and keyboard shortcuts, which help make your text editing faster and more efficient.

For future reference, Table 6.1 shows the symbols for the various keys on a typical Macintosh keyboard. Apply your technical sophistication (Box 5.2) if your keyboard differs.

Table 6.1: Miscellaneous keyboard symbols.

Key	Symbol
Command	⌘
Control	^
Shift	⇧
Option	⌥
Up, down, left, right	↑ ↓ ← →
Enter/Return	↵
Tab	→|
Delete	⌫

6.1 Choosing a Text Editor

While cloud IDEs have many advantages, every aspiring computer magician should learn at least one native editor (i.e., an editor that runs on your desktop operating system). While there are many editors to choose from, probably the most promising modern text editors in use today are Sublime Text (sometimes just called "Sublime"), Visual Studio Code (VScode), and Atom.[1] Each has its own advantages and disadvantages.

6.1.1 Sublime Text

- Advantages
    1. Powerful, hackable, and easy to use
    2. Can be used for free in "evaluation mode"
    3. Fast and robust, even when editing huge files/projects
    4. Works cross-platform (Windows, macOS, Linux)
    5. Backed by a profitable company that has a good track record of support and development

- Disadvantages
    1. Free in neither the speech nor beer sense

---

1. Other options include TextMate, NotePad++, jEdit, and BBEdit.

2. Has a mildly annoying popup that goes away only if you buy a license

3. Costs $70 as of this writing

4. Setting up command-line tools takes some fiddling

6.1.2   Visual Studio Code (VSCode)

- Advantages
    1. Powerful, with lots of packages

    2. Free to use

    3. Fast and robust, even when editing huge files/projects

    4. Works cross-platform (Windows, macOS, Linux)

    5. Backed by Microsoft

- Disadvantages
    1. Not open-source

    2. Backed by Microsoft

6.1.3   Atom

- Advantages
    1. Powerful, hackable, and easy to use

    2. Free in both the speech and beer senses (i.e., it both costs nothing and is open-source software)

    3. Works cross-platform (Windows, macOS, Linux)

    4. Easy to set up command-line tools

    5. Backed by collaboration powerhouse GitHub (https://github.com/)

- Disadvantages
    1. Reports of being slower in some cases than Sublime or VSCode

    2. Since GitHub's acquisition by Microsoft, backed by Microsoft

It's hard to go wrong with any of these choices. My main day-to-day editor as of this writing is Sublime Text, and I've heard great things about VSCode, but because of its simplicity and being 100% free I think Atom is probably the best choice for new users. The good news is that the skills in the sections that follow are near-universal;

if you learn Atom but decide to switch to Sublime Text or VSCode (or even a cloud IDE) for your daily editing, most of the core ideas will translate easily.

6.1.4 Exercises

Install and configure a text editor on your system as follows:

1. Download and install either Sublime Text, Visual Studio Code, or Atom.
2. If using Sublime Text, set up the **subl** command by Googling for "sublime text command line" and following the instructions for your system. Apply your technical sophistication (Box 5.2) if you get stuck. You might also find it helpful to skip ahead to Section 7.3 to learn about how to configure your system's *path*.
3. If using VSCode, set up the **code** command by Googling for "visual studio code command line" and following the instructions for your system.
4. If using Atom, go to Atom > Install Shell Commands to enable the **atom** command at the command line (Figure 6.1).

6.2 Opening

To open files, we're going to use the command configured in Section 6.1.4 to launch the editor and open the file at the same time (a method we used with **vim** in Section 5.3). In Section 7.4, we'll cover a second method (called "fuzzy opening") that's useful when editing a project with multiple files. I'll assume you're using the **atom** command, but if you're using Sublime you should make the appropriate substitution (**subl** in place of **atom**).

Let's get started by downloading a sample file, **README.md**, from the Web. As in Section 6.7.1 and Section 5.6, we'll use the **curl** command to download the file at the command line:

```
$ curl -OL https://cdn.learnenough.com/README.md
```

As hinted at by the **.md** extension, the downloaded file is written in Markdown, a human-readable markup language designed to be easy to convert to HTML, the language of the World Wide Web.

Figure 6.1: Installing Atom's shell commands.

After downloading **README.md**, we can open it at the command line as follows:

```
$ atom README.md
```

(If this doesn't work, be sure you've installed the Atom shell commands as shown in Figure 6.1.) The result of opening **README.md** in Atom should look something like Figure 6.2 or Figure 6.3. (If this is your first time opening Atom, it's also possible you'll see a one-time greeting screen. As usual, apply Box 5.2.) Figure 6.2 shows the usual default, which is for "word wrap" to be off; because Markdown files are typically written using a long line for each paragraph, this setting isn't ideal in this case, so I recommend turning on word wrap (called "soft wrap" in Atom) using the menu item shown in Figure 6.4.

In some editors, such as the cloud IDE at Cloud9, it's more common to open files using the filesystem navigator (although in fact the **c9** command can be used to open

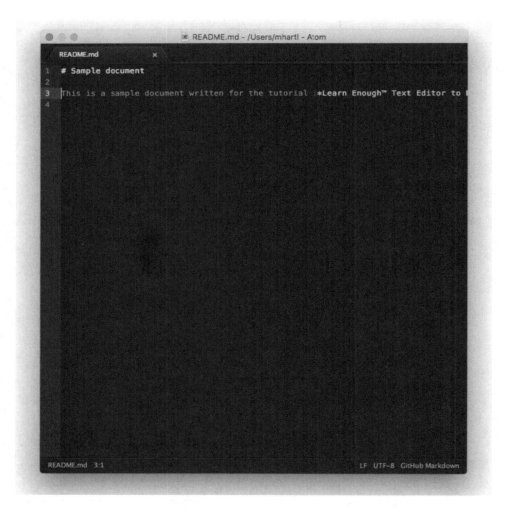

Figure 6.2: The sample file with word wrap off.

files at the Cloud9 command line).[2] Double-clicking on **README.md** in the filesystem navigator (Figure 6.5) opens the file in Cloud9's editor, as shown in Figure 6.6.

---

2. I used Cloud9 for over a year without discovering the **c9** command (which can be installed using **npm install -g c9** if it's not already present). Thanks to alert reader Timothy Kiefer for using his *technical sophistication* (Box 5.2) to figure it out!

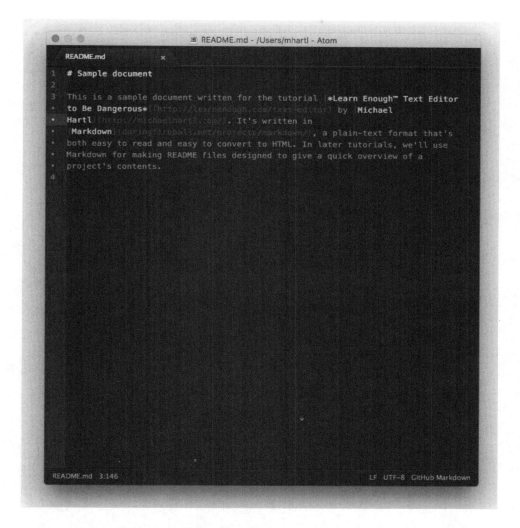

Figure 6.3: The sample file with word wrap on.

Figure 6.7 shows the file after clicking Navigate to close the filesystem navigator, and we see that, as in Figure 6.2, the line extends inconveniently off the screen. We can fix this using View > Wrap Lines as shown in Figure 6.8, with the word-wrapped result appearing as in Figure 6.9. (Figuring out that a menu item like View > Wrap

Figure 6.4: The menu item to toggle word wrap.

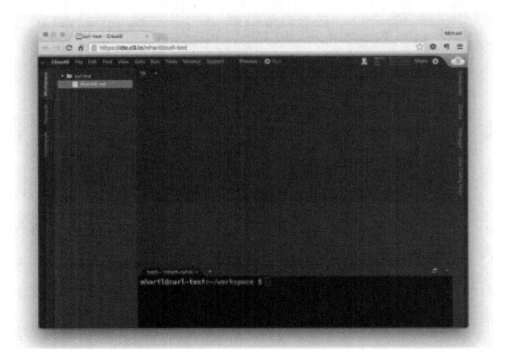

Figure 6.5: The Cloud9 filesystem navigator.

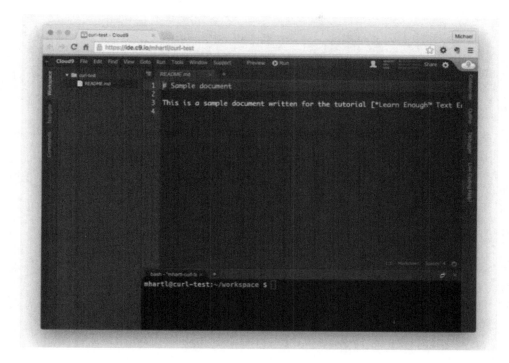

Figure 6.6: Cloud9 after double-clicking on **README.md**.

Lines turns on word wrap is exactly the kind of thing you should be able to figure out using your technical sophistication (Box 5.2).)

6.2.1 Syntax Highlighting

One thing you may have noticed from inspecting Figure 6.3 and Figure 6.9 is that both Atom and Cloud9 display different aspects of the file in different colors. For example, Atom shows characters inside square brackets **[]** (which represent text for HTML links) in a lighter color than the rest of the text, while Cloud9 shows the same text in green. This is a practice known as *syntax highlighting*, which makes special text formatting much easier to identify visually. It's essential to understand that this practice is strictly for our benefit; as far as the computer is concerned, the document in question is still plain text.

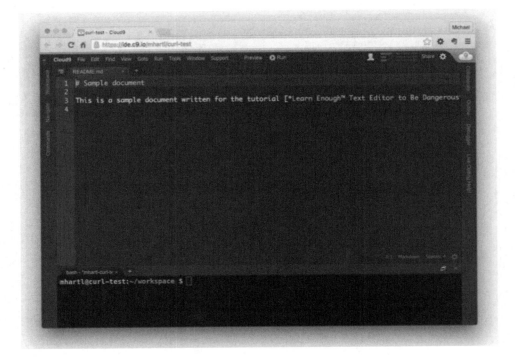

Figure 6.7: Cloud9 with word wrap off.

You might wonder how Atom and Cloud9 knew which highlighting scheme to use. The answer is that they infer the document format from the file type extension (in this case, `.md` for Markdown). The highlighting in Cloud9's case is quite high-contrast, but in Atom's case it isn't particularly prominent; the most significant things are the different colors for the heading

```
# Sample document
```

and for links like

```
[Michael Hartl](http://michaelhartl.com/)
```

We'll see more dramatic examples of syntax highlighting in Section 6.7 and especially in Section 7.2.

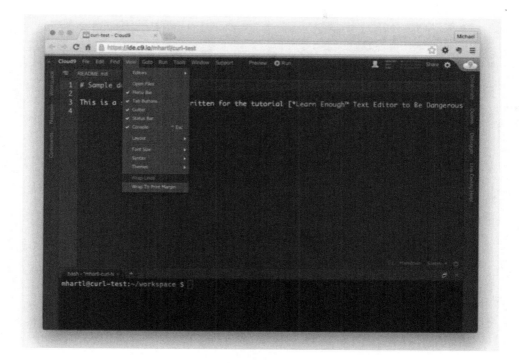

Figure 6.8: Turning word wrap on in Cloud9.

6.2.2   Previewing Markdown

As a final trick, I'd like to note that some editors, including Atom, can preview Markdown as HTML. We can figure out how to do this using our technical sophistication (Box 5.2), in this case by clicking on the Help menu and searching for "Preview" (Figure 6.10). The result is a built-in package called Markdown Preview, which converts Markdown to HTML and shows the result, as seen in Figure 6.11. In this context, it's convenient to work with an expanded width so that both the source and the preview are wide enough to view easily, as seen in Figure 6.12. This is accomplished by mousing over the side of the Atom window to get a double-arrow icon and then dragging to increase the size. We'll see another example of this "double-paned" setup in a more general setting starting in Section 7.4.

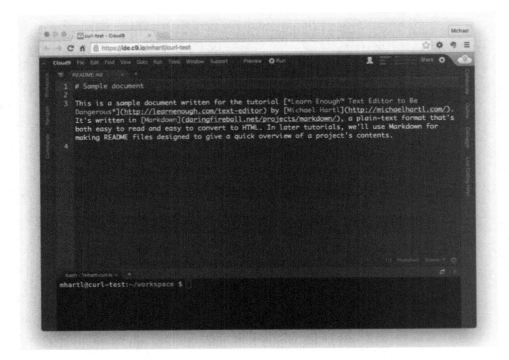

Figure 6.9: Cloud9 with word wrap on.

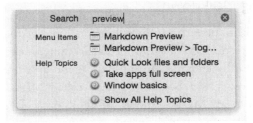

Figure 6.10: Using the Help menu to learn how to preview Markdown.

6.2.3 Exercises

1. By applying the methods in Box 5.2, find an online Markdown previewer (i.e., one that runs in a web browser), and use it to look at a preview of **README.md**. How do the results compare to Figure 6.11?

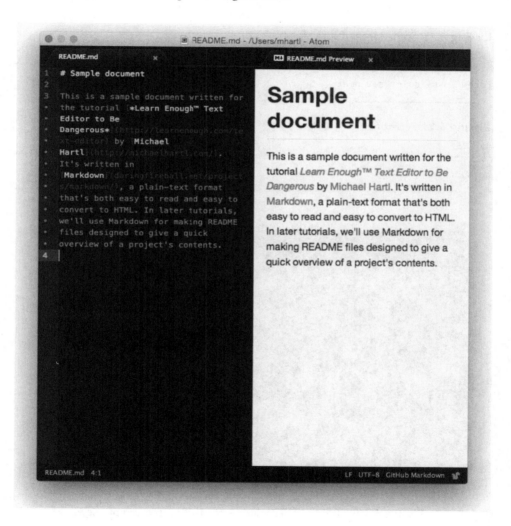

Figure 6.11: A Markdown preview in Atom.

2. Open a new document called `lorem.txt` and fill it with the text shown in Listing 6.1. Does the result have syntax highlighting?

3. Open a new document called `test.rb` and fill it with the text shown in Listing 6.2. Does the result have syntax highlighting?

Listing 6.1: Some lorem ipsum text.
~/lorem.txt

```
Lorem ipsum dolor sit amet
```

Listing 6.2: A test file.
~/test.rb

```
puts "test"
```

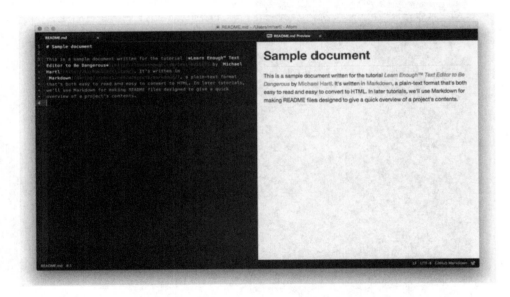

Figure 6.12: Using a wider window for the source and preview.

6.3 Moving

Unlike the commands for moving around in Vim (Chapter 5, summarized in Table 5.1), the commands for moving around in modern editors generally match the techniques used in other programs such as word processors, email programs, and web browsers. As a result, it's possible you may already know some or all of these techniques; if you don't, by following the steps in this section you'll get better at navigating other programs as a side effect.

To get started, let's open the large file from Section 5.6 consisting of the full text of Shakespeare's *Sonnets*:

```
$ atom sonnets.txt
```

(If this doesn't work, you may need to run the command in Listing 5.7, and you should also verify that you're in the right directory.) The result appears in Figure 6.13. Note that Figure 6.13 shows **sonnets.txt** in its own tab, with **README.md** from Section 6.2 occupying the other tab. Your result may vary; in any case, we'll discuss tabs further in Section 7.4.

As with most other native programs such as word processors, web browsers, etc., you can move around a modern editor using the mouse or trackpad. You can click to place the cursor, scroll using a scroll wheel or multi-touch gestures, or click and drag the scrollbar. The last of these is (as of this writing) incredibly subtle in Atom, so Figure 6.14 shows the scrollbar for Sublime Text. Figure 6.14 also shows the sort of two-pane view mentioned briefly in Figure 6.12, which we'll discuss more in Section 7.4.

In addition to using the mouse or trackpad, I also like using the arrow keys to move around, typically in concert with the Command key ⌘ (Table 1.1). (In Linux, Command is typically replaced with the Function key fn, and in Windows it's usually Ctrl, but you'll have to apply Box 5.2 to figure out the details.) My text editing typically involves lots of ⌘← and ⌘→ to move to the beginnings and ends of lines, and ⌘↑ and ⌘↓ to move to the top and bottom of the file. An example of moving to the end of the line in **README.md** appears in Figure 6.15, and an example of moving to the end of the file in **sonnets.txt** appears in Figure 6.16.

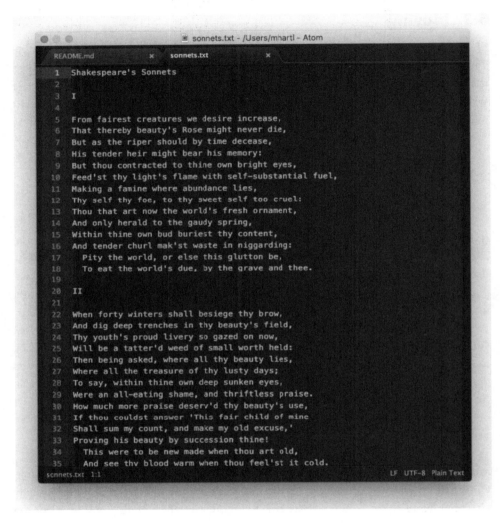

Figure 6.13: Opening Shakespeare's *Sonnets* in Atom.

6.3.1 Exercises

1. In your text editor, how do you move left and right one *word* at a time? *Hint*: On some systems, the Option key ⌥ might prove helpful.

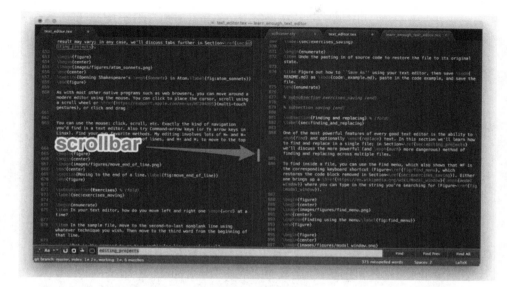

Figure 6.14: The Sublime Text scrollbar.

2. In **README.md**, move to the second-to-last nonblank line using whatever technique you wish. Then move to the third word from the beginning of that line.

3. What is the command to go to a particular line *number*? Use this command to go to line 293 of **sonnets.txt**. What do rough winds do?

4. By moving to the last nonblank line of **sonnets.txt** and pressing ⌘→ followed by ⌘←, show that ⌘← actually stops as soon as it reaches whitespace, with the result shown in Figure 6.17. How do you get to the true beginning of the line?

6.4   Selecting Text

Selecting text is an important skill that is particularly useful for deleting or replacing content, as well as for cutting, copying, and pasting (Section 6.5). Many of the techniques in this section make direct application of the commands to move around covered in Section 6.3. As in that section, the ideas here are quite general, applying to a wide variety of applications, not just to text editors.

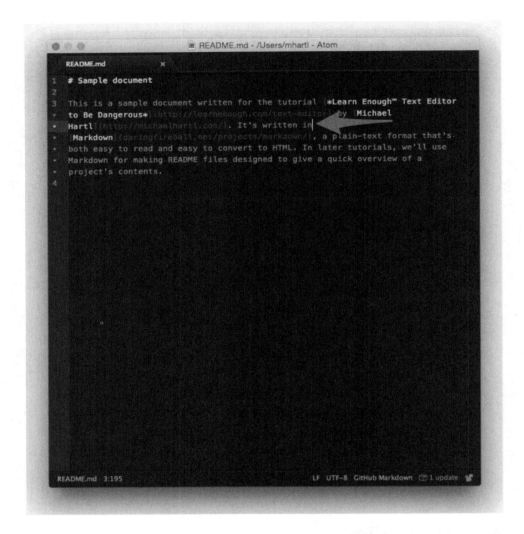

Figure 6.15: Moving to the end of a line with ⌘→.

In much the same way that modern editors make it easy to use the mouse to move the cursor, they also make it easy to use the mouse to select text. Simply click and drag the mouse cursor, as shown in Figure 6.18. Another closely related technique is to click on one location, and then Shift-click on another location to select all the text in between.

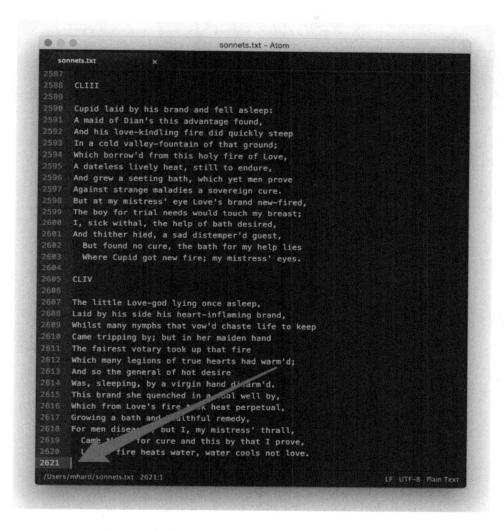

Figure 6.16: Moving to the end of the file with ⌘↓.

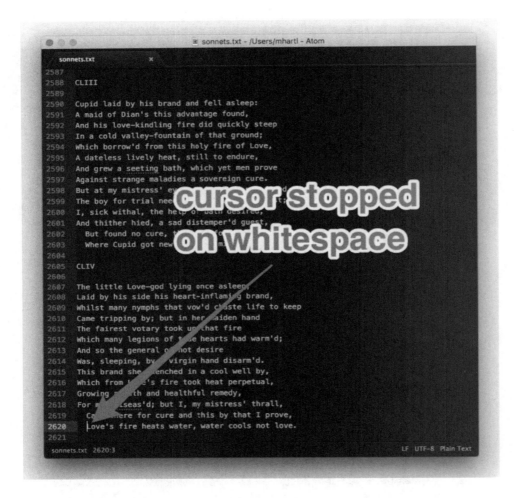

Figure 6.17: When using ⌘←, the cursor stops on whitespace.

6.4.1 Selecting a Single Word

When selecting text, there are some special cases that are useful enough to consider individually. We'll start with some techniques for selecting a single word:

- Click and drag the mouse cursor over the word.

- Double-click the word with the mouse.

- Press ⌘D (system-dependent; see Box 5.2).

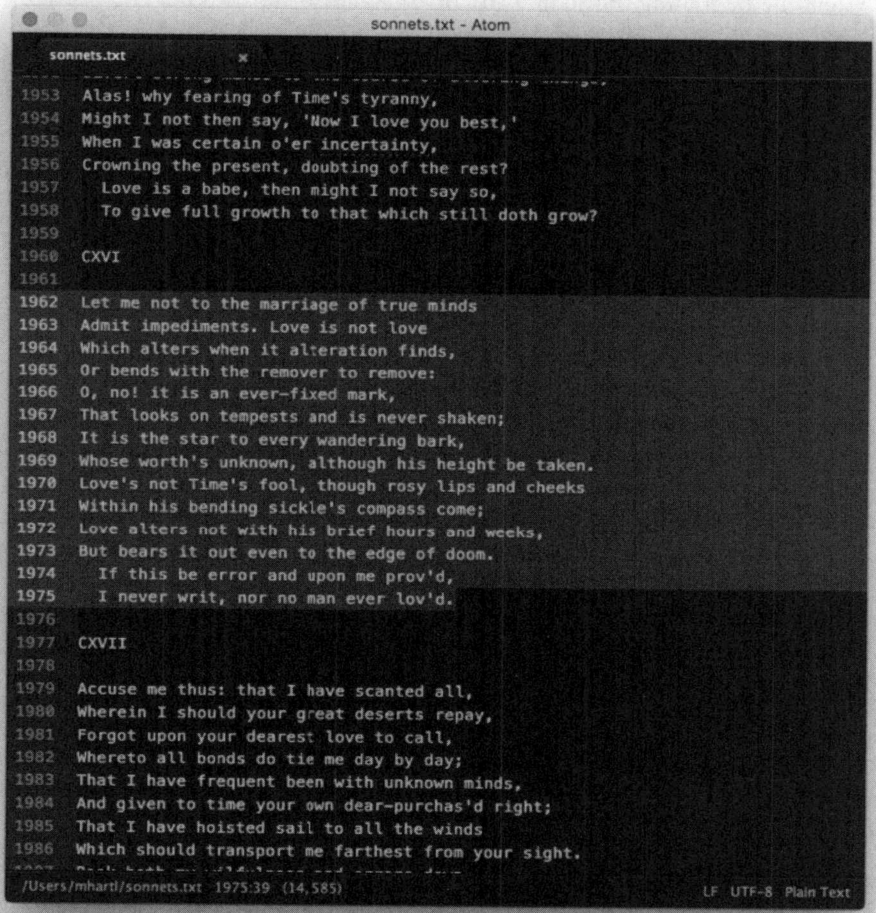

Figure 6.18: The result of clicking and dragging the mouse cursor.

6.4.2   Selecting a Single Line

Another technique, especially important when editing line-based text like computer code (or sonnets), involves selecting a full line or collection of lines. We start with ways to highlight a single line:

- Click the beginning of the line and drag the cursor to the end.

- Click the end of the line and drag the cursor to the beginning.

- Press ⌘← (twice) to get to the beginning of line, then press ⇧⌘→ to select to the end of line.

- Press ⌘→ to get to the end of line, then press ⇧⌘← (twice) to select to the beginning of line.

6.4.3 Selecting Multiple Lines

A comparably important technique is selecting multiple lines:

- Click and drag the mouse cursor over the words/lines.

- Hold down the Shift key and move the up- and down-arrow keys (⇧↑ and ⇧↓).

This latter technique is one of my personal favorites, and one of my most common editing tasks involves hitting ⌘← to go to the beginning of the first line I want to select and then hitting ⇧↓ repeatedly until I've selected all the lines I want (Figure 6.19). (As noted in Section 6.3.1, in many editors ⌘← stops on whitespace, so moving to the beginning of the line actually requires two uses of ⌘← in succession. Being able to figure out details and edge cases like this is a hallmark of growing technical sophistication (Box 5.2).)

6.4.4 Selecting the Entire Document

Finally, it's sometimes useful to be able to select the entire document at once. For this, there are two main techniques:

- Use a menu item called "Select All" or something similar. The specifics are editor-dependent; Figure 6.20 shows the use of the Selection menu in Sublime Text, while Figure 6.21 shows the use of the Edit menu in Atom.

- Press ⌘A.

Note from Figure 6.20 that the menu actually shows the corresponding command (⌘A); bootstrapping your knowledge using the menu items is a great way to learn keyboard shortcuts, which over time will make your text editing significantly more efficient.

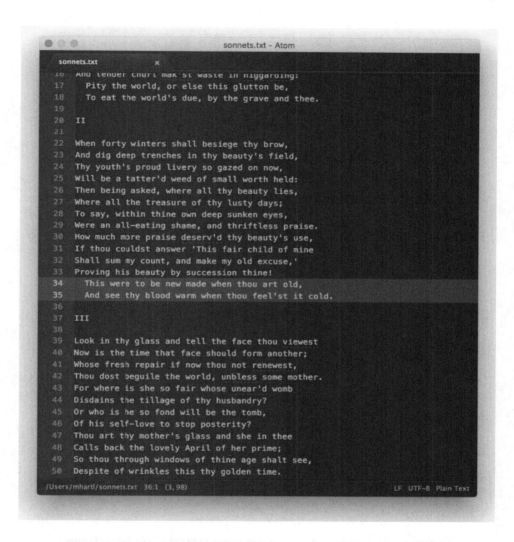

Figure 6.19: Selecting a Shakespearean couplet using ⌘← and ⇧↓.

6.4.5   Exercises

1. Select Shakespeare's second sonnet by clicking at the beginning and then Shift-clicking at the end.

Figure 6.20: Selecting the entire document using the Selection menu (Sublime Text).

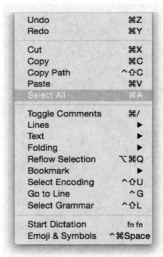

Figure 6.21: Selecting the entire document using the Edit menu (Atom).

2. Select the first line in the file by moving to the beginning with ⌘↑ and pressing ⇧⌘→ (or the equivalent for your system).

3. Delete the selection in the previous exercise (using the Delete key).

4. Select the word "document" in **README.md** and replace it with "README".

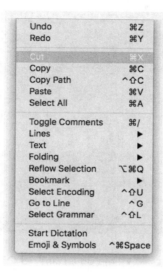

Figure 6.22: The Cut/Copy/Paste menu items (which you should never use).

6.5 Cut, Copy, Paste

The Cut/Copy/Paste triumvirate is one of the most useful sets of operations when editing text, especially when executed via the conveniently located keyboard shortcuts ⌘X/⌘C/⌘V. (Cut/Copy/Paste are available as menu items (Figure 6.22), but the operations are so common that I strongly recommend learning and using the keyboard shortcuts right away.) Although only ⌘C is mnemonic ("C" for "Copy"), the keys are conveniently located three in a row on the bottom row of a standard QWERTY keyboard, which makes it easy to use them in combination or in quick succession (Figure 6.23).

Applying either Cut or Copy involves first selecting text (Section 6.4), and then hitting either ⌘X to Cut or ⌘C to Copy. When using ⌘C to Copy, the selected text is placed in a *buffer* (temporary memory area); moving to the desired location (Section 6.3) and hitting ⌘V lets you Paste the content into the document at the location of the cursor. ⌘X works the same way as ⌘C, except the text is removed from the document as well as being copied into the buffer.

As a concrete example, let's select a Markdown link from the sample README file, **README.md**, as shown in Figure 6.24. After copying with ⌘C, we can then paste the

Figure 6.23: The XCV keys on a standard QWERTY keyboard.

link several times (with returns in between) by repeatedly hitting ⌘V and the Enter key, as shown in Figure 6.25. Finally, Figure 6.26 shows the result of cutting **README** from the main text and pasting it in at the end of the file.

6.5.1 Jumpcut

Although Cut/Copy/Paste is all that's strictly necessary for everyday editing, there is one big downside, which is that there is only room in the buffer for a single string. Among other things, this means that if you Cut something and then accidentally hit "copy" instead of "paste" (which is easy since the letters are adjacent on the keyboard), you overwrite the buffer, and the text you Cut is gone forever (unless you undo as described in Section 6.6). If you happen to be developing on a Mac, there's a solution to this problem: a free program called Jumpcut (https://snark.github.io/jumpcut/). This remarkable little utility app expands the buffer by maintaining more than one entry in the history. You can navigate this expanded buffer using either the Jumpcut menu (Figure 6.27) or the keyboard shortcuts ^⌥V (cycle forward) and ⇧^⌥V (cycle backward). I use Jumpcut dozens or even hundreds of times a day, and I strongly suggest giving it a try.

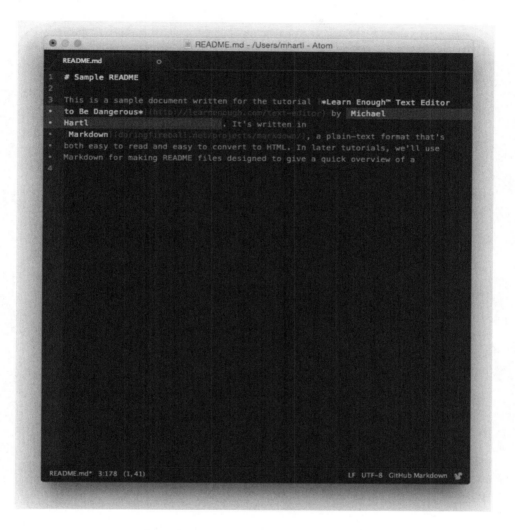

Figure 6.24: Selecting a Markdown link.

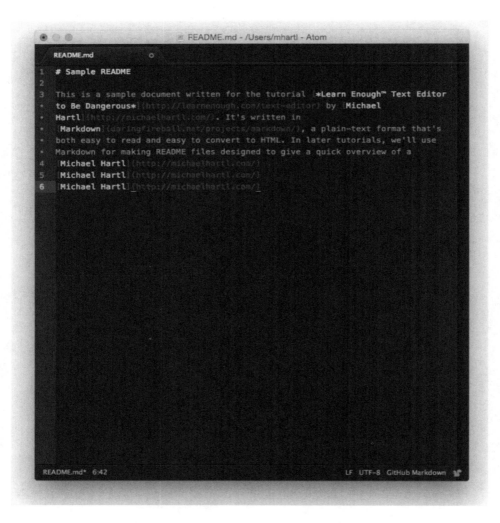

Figure 6.25: Pasting link text several times (with returns in between).

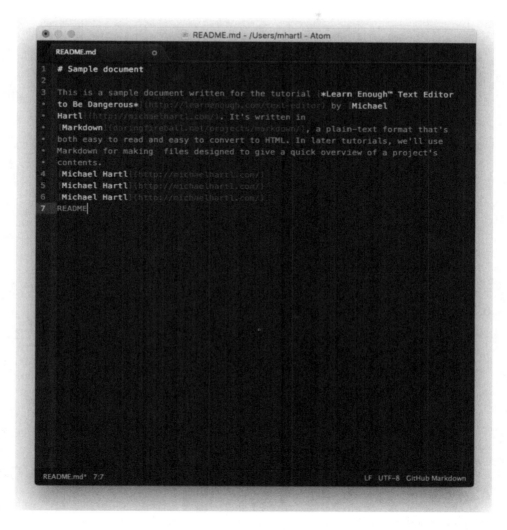

Figure 6.26: The result of cutting "README" and pasting at the end of the file.

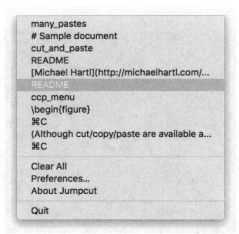

Figure 6.27: Jumpcut expands the copy-and-paste buffer to include a longer history.

6.5.2 Exercises

1. Select the entire document, Copy it, and Paste several times. The result should look something like Figure 6.28.

2. Select the entire document and Cut it. Why might this be preferable to deleting it?

3. Select and copy the couplet at the end of Sonnet 1 and paste it into a new file called **sonnet_1.txt**. How do you create a new file directly in your editor?

6.6 Deleting and Undoing

We mentioned deleting before in Section 6.4.5 (the exercises for Section 6.4), which of course simply involves pressing the Delete key, sometimes written as ⌫ (Table 1.1). As with Cut/Copy/Paste (Section 6.5), deletion is especially useful when combined with the selection techniques from Section 6.4.

In addition to the obvious technique of selecting and deleting text, on a Mac I especially like using ⌥⌫ to delete one word at a time. I'll frequently use this combination if I need to delete a medium number of words (say 2–5) to restart a phrase when writing. For shorter deletion tasks, such as one word, it's usually faster to

Figure 6.28: The result of pasting the whole document several times.

hit ⌫ repeatedly, as context-switching to use ⌥⌫ incurs some overhead that makes it faster to just delete directly. Don't worry too much about these micro-optimizations, though; with experience, as a matter of course you'll come up with your own set of favorite techniques.

Figure 6.29: Undo and Redo in the editor menu.

Paired with deletion is one of the most important commands in the history of the Universe, Undo. In modern editors, Undo uses the native keybinding, typically ⌘Z or ^Z. Its inverse, Redo, is usually something like ⇧⌘Z or ⌘Y. You can also use the menu (typically Edit, as seen in Figure 6.29), but, as with Cut/Copy/Paste (Section 6.5), Undo is so useful that I recommend memorizing the shortcut as soon as possible. Without Undo, operations like deletion would be irreversible and hence potentially harmful, but with Undo it's easy to reverse any mistakes you make while editing.

One practice I recommend is using Cut instead of Delete whenever you're not 100% sure you'll never want the content again. Although you can usually Undo your way to safety if you accidentally delete something important, putting the content into the buffer with Cut gives you an additional layer of redundancy. (Using Jumpcut (Section 6.5.1) gives you another layer still.)

Finally, Undo provides us with a useful trick for finding the cursor, a common task when editing larger files. The issue is that you'll be writing some text and then need to move (Section 6.3) or find (Section 6.8) elsewhere in the document. On these occasions, it can be hard to relocate the cursor. There are several ways around this problem—you can move the arrow keys, or just start typing—but my favorite technique is to Undo and then immediately Redo (⌘Z/⇧⌘Z or ⌘Z/⌘Y), which is guaranteed to find the cursor without making any undesired changes.

6.6.1   Exercises

1. Use Undo repeatedly until all the changes you've made to **README.md** have been undone.

2. Using any technique you want from Section 6.4, select the word "written" in **README.md** and delete it, then undo the change.

3. Redo the change from the previous exercise, then undo it again.

4. Make an edit somewhere in **sonnets.txt**, then scroll around so you get lost. Use the Undo/Redo trick to find the cursor again. Then keep using Undo to undo all your changes.

6.7 Saving

Once we've made some edits to a file, we can save it using the menu or with ⌘S. I strongly recommend using the keyboard shortcut, which among other things makes it easier to save the file whenever you reach a temporary pause in your writing or coding—a valuable habit to cultivate. Basically, if you're not doing something else, you should be hitting Save. This habit goes a long way toward preventing lost work (and, as discussed in Part III, is especially powerful when combined with version control).

As an example, we can add some source code to our README file and save the result. We start by pasting in the code from Listing 6.3, as shown in Figure 6.30 (which includes some nice high-contrast syntax highlighting). As you can see from the circled indicator in Figure 6.30, Atom (as with most modern editors) includes a subtle indicator that the file is unsaved, in this case a small open circle. After running Save (via ⌘S, for example), the circle disappears, to be replaced with an X (Figure 6.31).

Listing 6.3: A code snippet.

```ruby
```ruby
def hello
  puts "hello, world!"
end
```
```

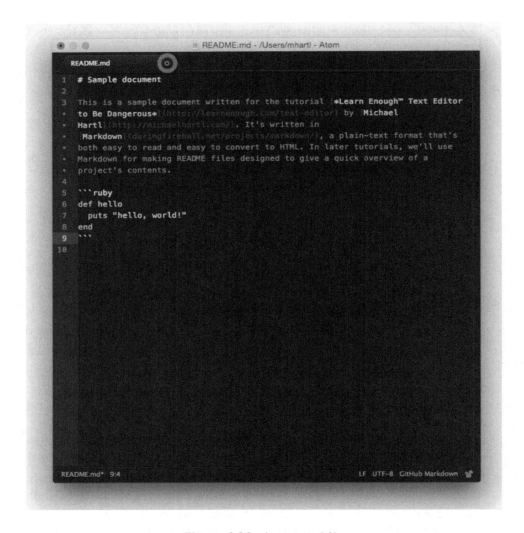

Figure 6.30: An unsaved file.

6.7.1 Exercises

1. Undo the pasting in of source code to restore the file to its original state.

2. Figure out how to "Save As", then save **README.md** as **code_example.md**, paste in the code example, and save the file.

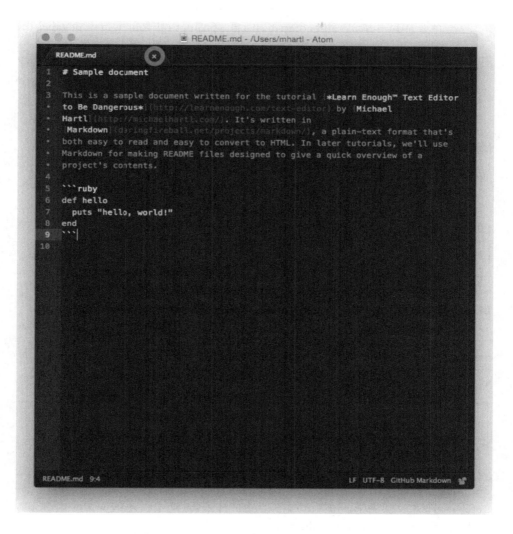

Figure 6.31: The file from Figure 6.30 after saving.

3. The default Bash prompt for my command-line terminal appears as in Listing 6.4, but I prefer the more compact prompt shown in Listing 6.5. In Part I (Section 4.3), I promised to show how to customize the prompt in Part II. Fulfill this promise by editing the `.bashrc` file to include the lines shown in Listing 6.6. Source the Bash

profile as in Listing 5.5 and confirm that the prompt on your system matches the one shown in Listing 6.5. (To learn how to customize the prompt using Z shell, the current default shell on macOS, see the Learn Enough blog post "Using Z Shell on Macs with the Learn Enough Tutorials" (https://news.learnenough.com /macos-bash-zshell).)

Listing 6.4: The default terminal prompt on my system.

```
MacBook-Air:~ mhartl$
```

Listing 6.5: My preferred, more compact prompt.

```
[~]$
```

Listing 6.6: The Bash lines needed to customize the prompt as shown in Listing 6.5.
~/.bashrc

```
alias lr='ls -hartl'
# Customize prompt to show only working directory.
PS1='[\W]\$ '
```

6.8 Finding and Replacing

One of the most powerful features of every good text editor is the ability to *find* and optionally *replace* text. In this section we'll learn how to find and replace in a single file; in Section 7.4 we'll discuss the more powerful (and *much* more dangerous) method of finding and replacing across multiple files.

To find inside a file, you can use the Find menu, shown in Figure 6.32, which also shows that ⌘F is the corresponding keyboard shortcut. Either one brings up a *modal window* where you can type in the string you're searching for (Figure 6.33).

For example, suppose we search for the string "sample". As seen in Figure 6.34, both "Sample" and "sample" are highlighted. The reason our search finds both is because we've opted to search case-insensitively (which is usually the default).

Figure 6.32: Finding using the menu.

Figure 6.33: A modal window for finding and replacing.

Figure 6.35 shows how to use the modal window to find "sample" and replace with "example". In order to avoid replacing "Sample", we first click on Find to select the next match, and then click on Replace to replace the second match (Figure 6.36). (Changing to case-sensitive search would also work in this case; learning how to do this is left as an exercise (Section 6.8.1).)

As seen in Figure 6.32, you can also type ⌘G to find the next match using a keyboard shortcut. This ⌘F/⌘G combination also works in many other applications, such as word processors and web browsers.

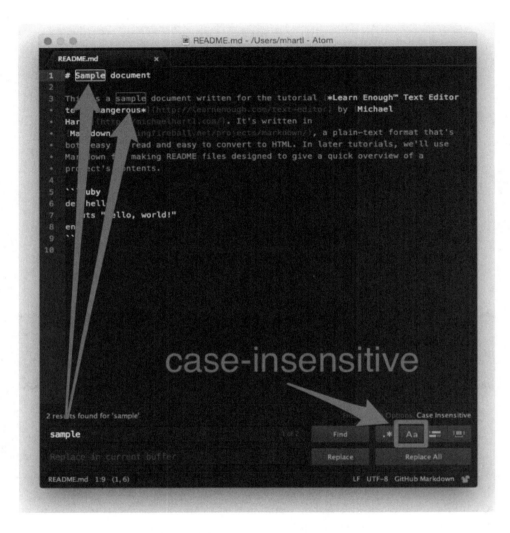

Figure 6.34: Finding the string "sample".

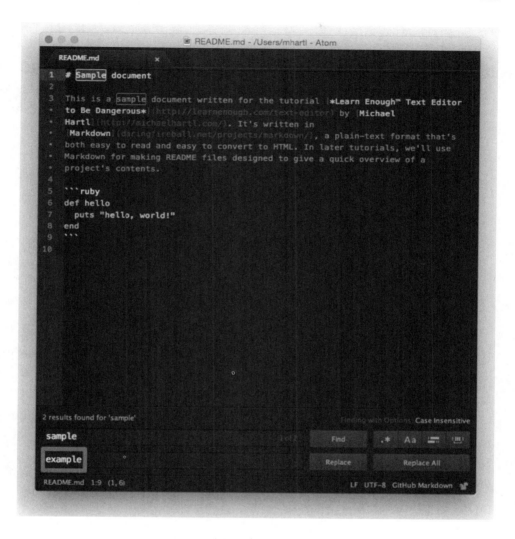

Figure 6.35: Finding and replacing.

Figure 6.36: The result of replacing "sample" with "example".

6.8.1 Exercises

1. In Section 6.3.1, we found Sonnet 18 by going directly to line 293, but of course I didn't search the file line by line to write the exercise. Instead, I searched for "shall I compare thee". Use your text editor to search for this string in **sonnets.txt**. On what line does "rosy lips and cheeks" appear?

2. The example in this section shows one of the pitfalls of mechanically finding and replacing text: We've ended up with the ungrammatical result "a example" instead of "an example". Rather than fix this by hand, use find and replace to replace "a example" with "an example" in your document. (Although in the present case there's only one occurrence, this more general technique scales up to documents much longer than our toy example.)

3. What is the keyboard shortcut in your editor for finding the previous match?

4. What is the keyboard shortcut to replace in the current buffer (file)? How does this differ from the keyboard shortcut for simply finding?

6.9 Summary

- Atom, Sublime Text, and VSCode are all excellent choices for a primary modern text editor.

- One common way to open files is to use a command at the command line.

- For files containing things like prose with long lines, it's a good idea to turn on word wrap.

- Moving around text files can be accomplished many different ways, including using the mouse and arrow keys (especially in combination with the Command/Control key).

- One convenient way to select text is to hold down Shift and move the cursor.

- The Cut/Copy/Paste triumvirate is incredibly useful.

- Undo can save your bacon (Figure 6.37).[3]

Important commands from this chapter are summarized in Table 6.2.

---

3. Image courtesy of peter s./Shutterstock.

Figure 6.37: Undo can save your bacon.

Table 6.2: Important commands from Chapter 6.

Command	Description
⌘←	Move to beginning of line (stops on whitespace)
⌘→	Move to end of line
⌘↑	Move to beginning of file
⌘↓	Move to end of file
⇧-move	Select text
⌘D	Select current word
⌘A	Select All (entire document)
⌘X/⌘C/⌘V	Cut/Copy/Paste
⌘Z	Undo
⇧⌘Z or ⌘Y	Redo
⌘S	Save
⌘F	Find
⌘G	Find next

CHAPTER 7

Advanced Text Editing

Having covered the basic functions of modern text editors in Chapter 6, in this chapter we'll learn about a few of the most common advanced topics. Even more than in Chapter 6, details will vary based on the exact editor you choose, so use your growing technical sophistication (Box 5.2) to figure out any necessary details. The most important lesson is that the advanced functions in this chapter are all things that *any* professional-grade editor can do, so you should be able to figure out how to do them no matter which editor you're using.

7.1 Autocomplete and Tab Triggers

Two of the most useful features of text editors are *autocomplete* and *tab triggers*, which you can think of as roughly command-line style tab completion for text files. (See Box 2.4 from Part I for details on tab completion.) Both features allow us to type potentially large amounts of text with only a few keystrokes.

7.1.1 Autocomplete

The most common variant of autocomplete lets us type the first few letters of a word and then gives us the ability to complete it from a menu of options, typically by using the arrow keys and hitting tab to accept the completion. An example of autocompleting the word "Markdown" in **README.md** appears in Figure 7.1.

The autocomplete menu itself is populated using the current document, so autocomplete is particularly useful in longer documents that contain a large number of

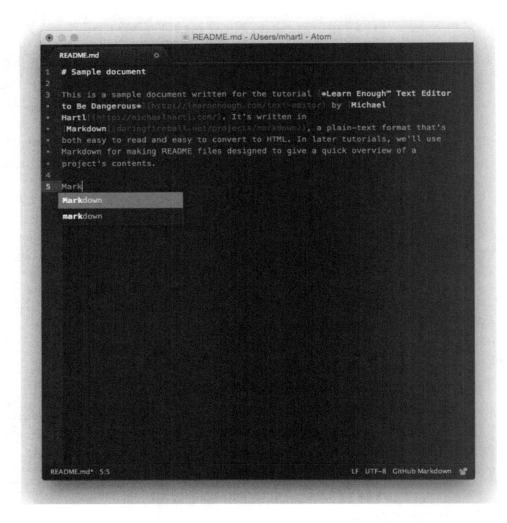

Figure 7.1: Autocomplete for "Markdown".

possible completions. For instance, the source for *Learn Enough Text Editor to Be Dangerous* (https://www.learnenough.com/text-editor) (which is written using the powerful markup language LaTeX) makes use of a large number of labels for making cross-references, and these labels are often long enough that it's much easier to autocomplete them than to type them out by hand. An example is the oft-cited Box 5.2, whose source looks like Listing 7.1.

Listing 7.1: A cross-reference with a label I usually autocomplete.

```
Box~\ref{aside:technical_sophistication_text_editor}
```

When writing a string like **technical_sophistication** in Listing 7.1, I nearly always use autocomplete instead of typing it out in full.[1] (As mentioned below, the rest of the cross-reference is generated using a custom tab trigger.) Similar considerations frequently occur when writing source code, where (as we'll learn in *Learn Enough Ruby to Be Dangerous* (https://www.learnenough.com/ruby)) we might encounter something like this:

```
ReallyLongClassName < ReallyLongBaseClassName
```

In such cases, rather than typing out the long names by hand, it's usually easier to type **Rea** and then select the relevant autocompletion.

7.1.2 Tab Triggers

Tab triggers are similar to autocompletion in that they let us type a few letters and then hit *Tab* to work some magic, but in this case many of them come pre-defined with the editor, with the exact triggers typically based on the particular type of document we're editing. For example, in Markdown and other markup files (HTML, LaTeX, etc.), typing `lorem→|` or `lo→|` yields so-called *lorem ipsum* text, a slightly corrupted Latin fragment from a book by Cicero that is often used as dummy text in programming and design. We saw *lorem ipsum* briefly before in Listing 6.1; a second example appears in Figure 7.2, which shows the result of typing `lo` in Atom. A closeup appears in Figure 7.3. After hitting `→|` to invoke the tab trigger, the full *lorem ipsum* text appears as in Figure 7.4.

---

1. I actually have my Sublime Text editor configured to use the **ESC** key for autocompletion instead of using a menu, mainly because I got used to that design when using my previous editor (TextMate). I arranged for this setup using my technical sophistication (Box 5.2).

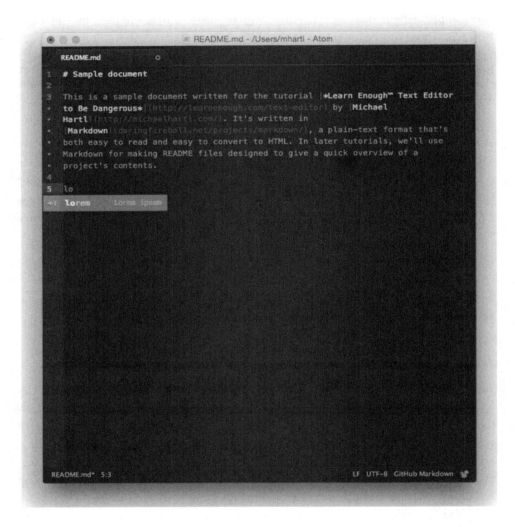

Figure 7.2: Typing "lo" in Atom prepares to activate a tab trigger.

Tab triggers are especially useful when editing more syntax-heavy file types like HTML and source code. For instance, when writing HTML, many editors support the creation of an HTML skeleton using the trigger html→|, together with HTML *tags* (covered in *Learn Enough HTML to Be Dangerous*

Figure 7.3: A more detailed view of the trigger in Figure 7.2.

(https://www.learnenough.com/html)) using the tag name with a tab, such as h1 →|
for an **h1** or top-level heading tag. In Atom, we can do something like this:

```
$ atom index.html
```

The result of applying the various tab triggers then might look something like
Figure 7.5. Because HTML, or HyperText Markup Language, is the language of the
World Wide Web, navigating to the file in a browser then shows a simple but real web
page (Figure 7.6).

Similarly, when writing Ruby code, typing def →| in Atom creates a Ruby *define*
statement to make a *function*, which looks like this:

```
def method_name

end
```

After typing the name of the function (which replaces the placeholder text
method_name), we can hit →| again to place the cursor in the right location to start
writing the main part of the function. These sorts of auto-expansions of content can
speed up code production considerably, while also lowering the cognitive load of
programming. We'll see a concrete example of this technique in Section 7.2.

Finally, it's possible to define tab triggers of your own. My own editing makes
extensive use of tab triggers; for example, to make the text in Listing 7.1, instead of
typing

```
Box~\ref{aside:technical_sophistication_text_editor}
```

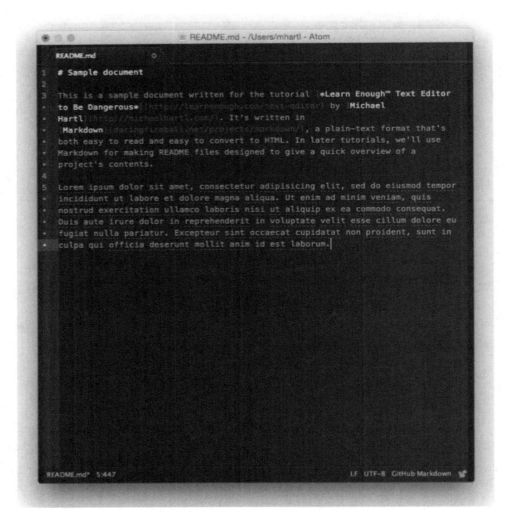

Figure 7.4: The result of the tab trigger in Figure 7.2.

by hand I used the custom tab trigger **bref** (for "box reference") to generate

```
Box~\ref{aside:}
```

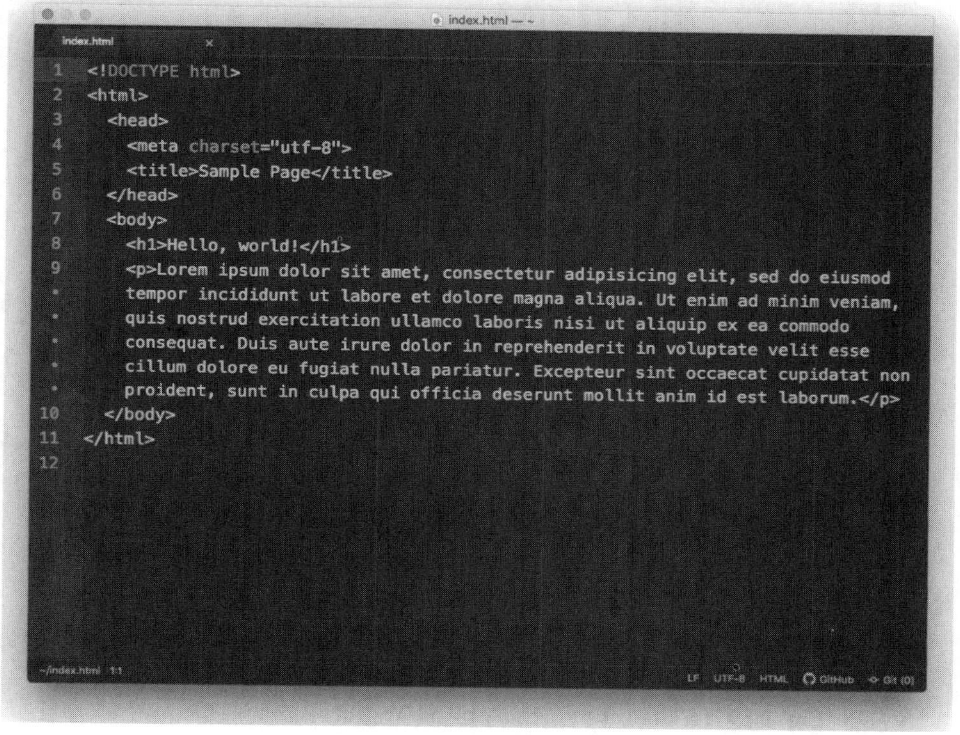

```
index.html — ~
index.html                    x
1   <!DOCTYPE html>
2   <html>
3     <head>
4       <meta charset="utf-8">
5       <title>Sample Page</title>
6     </head>
7     <body>
8       <h1>Hello, world!</h1>
9       <p>Lorem ipsum dolor sit amet, consectetur adipisicing elit, sed do eiusmod
        tempor incididunt ut labore et dolore magna aliqua. Ut enim ad minim veniam,
        quis nostrud exercitation ullamco laboris nisi ut aliquip ex ea commodo
        consequat. Duis aute irure dolor in reprehenderit in voluptate velit esse
        cillum dolore eu fugiat nulla pariatur. Excepteur sint occaecat cupidatat non
        proident, sunt in culpa qui officia deserunt mollit anim id est laborum.</p>
10    </body>
11  </html>
12
```

Figure 7.5: The result of applying HTML tab triggers.

and then filled in the label `technical_sophistication` using autocomplete
(Section 7.1.1). Defining custom tab triggers is highly editor-dependent and is beyond
the scope of this tutorial, but some hints about how to figure it out for yourself appear
in Section 7.5.

7.1.3   Exercises

1. Add some more *lorem ipsum* text to **README.md** using a tab trigger.

2. Add another occurrence of the word "consectetur" using autocomplete.

3. Write the sentence "As Cicero once said, 'quis nostrud exercitation ullamco
laboris'." with the help of as many uses of autocomplete as you want.

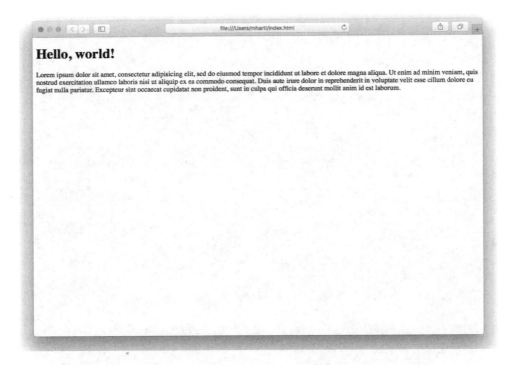

Figure 7.6: The result of applying tab triggers to an HTML page.

7.2 Writing Source Code

As hinted at in Section 7.1.2, in addition to being good at editing markup like HTML and Markdown, text editors excel at writing computer programs. Any good programmer's text editor supports many specialized functions for writing code; this section covers a few of the most useful. Even if you don't know how to program (yet!), it's still useful to know about some of the ways text editors support writing code.

An example of computer code appears in Listing 7.2, which shows a variant of a "hello, world" program written in the Ruby programming language. (You are not expected to understand this program.)

Listing 7.2: A variant of "hello, world" in Ruby.

```
1  # Prints a greeting.
2  def hello(location)
3    puts "hello, #{location}!"
4  end
5
6  hello("world")
```

To see the contents from Listing 7.2 in a text editor, we can fire up Atom as follows:

```
$ atom hello.rb
```

Upon pasting in the content of Listing 7.2, we get the result shown in Figure 7.7. (For extra credit, type in Listing 7.2 by hand using the **def** tab trigger discussed in Section 7.1.)

7.2.1 Syntax Highlighting

As we saw in Section 6.2.1 with **README.md**, Atom uses the filename extension to determine the proper syntax highlighting. In that case the (rather subtle) highlighting was for Markdown; in this case, Atom infers from **.rb** that the file contains Ruby code, and highlights it accordingly. As before, it's essential to understand that the highlighting isn't inherent to the text, which is still plain. Syntax highlighting is purely for our benefit as readers of the code.

In addition to making it easier to parse the source code visually (e.g., distinguishing keywords, strings, constants, etc.), syntax highlighting can also be useful for catching bugs. For example, at one point when editing *Learn Enough Command Line to Be Dangerous* (https://www.learnenough.com/command-line) I accidentally deleted a LATEX closing quote (which consists of the two single quotes ' '), with the result shown in Figure 7.8. This changed the color of the main text from the default white to the color used for quoted strings (green), which made it apparent at a glance that something was wrong. Upon fixing the error, the highlighting changed back to the expected white text, as shown in Figure 7.9.

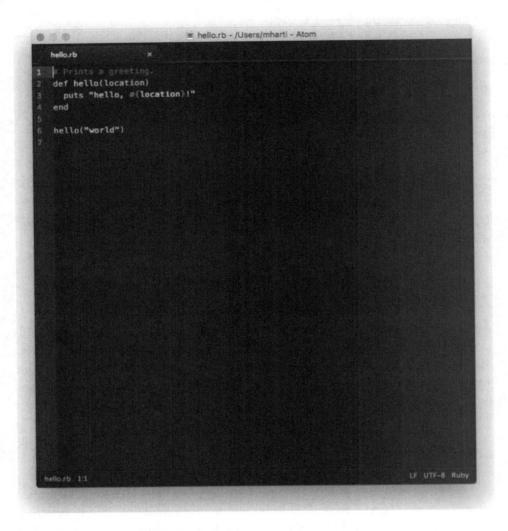

Figure 7.7: A Ruby program in Atom.

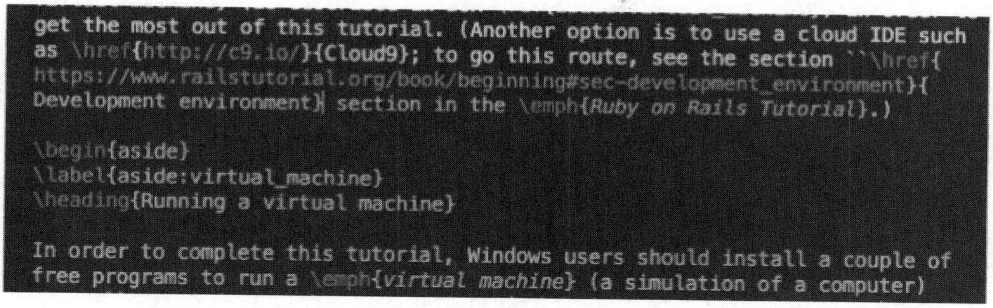

Figure 7.8: An error in LaTeX source caught by syntax highlighting.

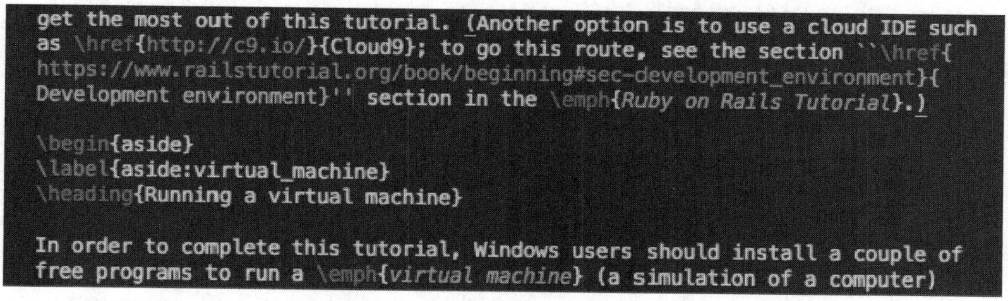

Figure 7.9: Error fixed, syntax highlighting as expected.

7.2.2 Commenting Out

One of the most useful functions of a text editor is the ability to "comment out" blocks of code, a technique often used to temporarily prevent execution of certain lines without having to delete them entirely (which is often particularly helpful when debugging). Most programming and markup languages support comment lines that exist for the benefit of humans reading the code but are ignored by the programming language itself.[2] An example of a Ruby comment appears in the first line of Listing 7.2:

```
# Prints a greeting.
```

Here the leading hash symbol # is Ruby's way of indicating a comment line.

---

2. Technically, comments are ignored by the compiler or interpreter. Some languages have automated documentation systems that do process the comments.

Suppose we wanted to comment out the next three lines (lines 2–4), to change

```
# Prints a greeting.
def hello(location)
  puts "hello, #{location}!"
end

hello("world")
```

to

```
# Prints a greeting.
# def hello(location)
#   puts "hello, #{location}!"
# end

hello("world")
```

It's possible to do this by hand, of course, simply by inserting a # at the beginning of each line. This is inconvenient, though, and becomes increasingly so as the length of the commented-out text grows. Instead, we can select the desired text (Section 6.4) and use a menu item or keyboard shortcut to comment out the selection. In Atom, we can comment out lines 2–3 by selecting those lines (Figure 7.10) and hitting ⌘/, as shown in Figure 7.11. (Note from Figure 7.11 that the subtle save indicator shown in Figure 6.31 has been filled in; this is because I habitually press ⌘S after making changes, as recommended in Section 6.7.)

The commenting-out feature typically toggles back and forth, so by hitting ⌘/ a second time we can restore the file to its previous state (Figure 7.10). This is useful when restoring some commented-out text after, for example, doing some debugging.

7.2.3 Indenting and Dedenting

Another element of code formatting made easier by text editors is *indentation*, which consists of the leading spaces at the beginning of certain lines. It used to be common to use *tab* characters for indentation, but unfortunately the number of spaces *displayed* for a tab is system-dependent, leading to unpredictable results: Some people might see four "spaces" per tab, some might see eight, and some might see only two.

In recent years, many programmers have switched to *emulated tabs*, where pressing the tab key inserts a standard number of ordinary spaces (typically two or four). True

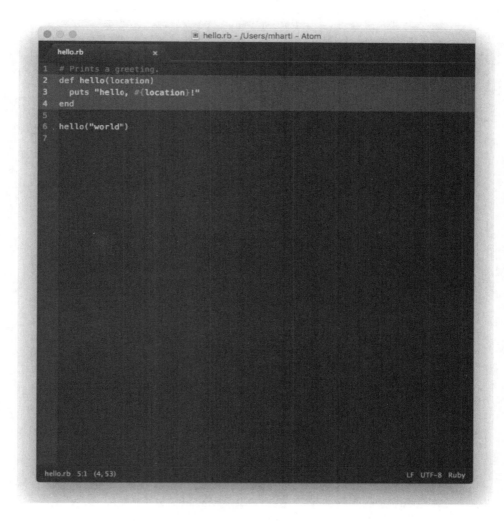

Figure 7.10: Preparing to comment out some lines.

tabs still have some partisans, though, and tabs vs. spaces remains holy war territory (Box 5.4). (Luckily, there is one thing everyone agrees on, which is that *mixing* tabs and spaces is a bad idea.)

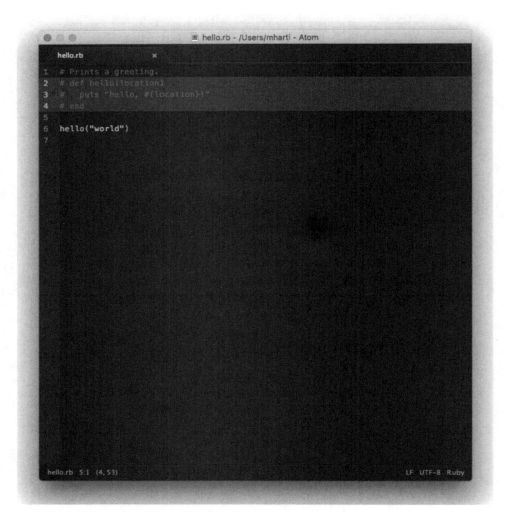

Figure 7.11: Commented-out lines.

To see how this works, we can take a look at some Ruby code, which typically uses two spaces for indentation:

```
def hello(location)
  puts "hello, #{location}!"
end
```

This would typically be achieved by hitting return after "(location)" and then pressing the tab key, although pressing the spacebar twice would also work. Assuming that the editor has been configured to use two spaces to emulate tabs, we'd get the result shown above. In most languages, this would be equivalent to the following:[3]

```
def hello(location)
puts "hello, #{location}!"
end
```

This second example is harder to read, though, and it's important to indent properly for the sake of humans reading the code, even if the programming language doesn't care.[4]

Text editors help maintain proper indentation in two main ways. First, new lines are typically inserted at the same level of indentation as the previous line, which you can verify by going to the end of line 3 in Listing 7.2 and typing in the following two lines:

```
puts "Uh, oh."
puts "Goodbye, #{location}!"
```

The result appears in Figure 7.12.

The second main way text editors help maintain good code formatting is by supporting block indentation, which works in much the same way as commenting out code blocks. Suppose, for example, that (contrary to conventional Ruby practices) we decided to indent lines 3–5 in Figure 7.12 six extra spaces, making eight spaces total. As with commenting out, the first step is to select the text we want to indent (Figure 7.13). We can then type the tab key →| to indent one "soft tab" (which

---

3. Python is a notable exception.

4. Some languages, notably Python, actually enforce some measure of proper indentation, but most language compilers and interpreters ignore it.

Figure 7.12: Adding two indented lines.

is usually two spaces for Ruby) at a time. (If for any reason the default indentation in your editor doesn't match the convention for the language you're using, apply your technical sophistication (Box 5.2) to figure out how to change it.) The result of applying three tabs in succession is shown in Figure 7.14.

Figure 7.13: Preparing to indent some lines.

Figure 7.14: A block of Ruby code indented more than usual.

Figure 7.15: Dedenting the code block in Figure 7.14.

Because each extra tab just indents the block more, we can't use the same command to undo indentation the way we did when commenting out code. Instead, we need to use a separate "dedent" command, which in Atom is ⇧ ⭲. Applying this command three times in succession returns us to our original state, as shown

Figure 7.16: The modal box for going to a particular line number.

in Figure 7.15. (By the way, many editors (including Atom) support the alternate keyboard shortcuts ⌘] and ⌘[for indenting and dedenting, respectively.)

7.2.4 Goto Line Number

It's often important to be able to go to a particular line number, such as when debugging a program that has an error on (say) line 187. We saw this feature in Section 5.6, where we learned that the Vim command **<n>G** takes us to line **<n>**. In many other editors, the relevant shortcut is ^G. This opens a modal box where you can type in the line number, as shown in Figure 7.16. (Incidentally, the syntax **:<number>** shown in Figure 7.16, which is for Sublime Text, also works in Vim.)

7.2.5 80 Columns

Finally, many text editors help programmers enforce a line limit of 80 characters across, usually called an "80-column limit". Not all programmers observe this limit, but keeping our code to 80 columns makes it easier to read and display, for example in fixed-width terminals, blog posts, or tutorials such as this one.[5] An 80-column limit also enforces good coding discipline, as exceeding 80 columns is often a sign that we would do well to introduce a new variable or function name.[6] Because it's difficult to tell at a glance if a particular line exceeds 80 characters, many editors (including Atom and Sublime Text) include the option to display a subtle vertical line showing where the limit is, as shown in Figure 7.17.[7] If the 80-column limit indicator isn't shown by

---

5. The original source of this constraint is actually IBM punch cards.

6. The main exception to the 80-column rule is markup like HTML, Markdown, or LaTeX, which is why in these cases we often activate word wrap as in Section 6.2.

7. In some editors the line is at something like 78 columns instead of 80 to allow a small margin for error.

Figure 7.17: Unsubtle arrows pointing at the subtle 80-column indicator in Atom.

default in your editor, flex your technical sophistication to figure out how to enable it. (It's often associated with a setting called "word wrap column".)

7.2.6 Exercises

1. Create the file **foo.rb**, then define the class **FooBar** (Listing 7.3) using a tab trigger. *Hint*: Chances are the trigger is something like cla→|.

2. Referring to Listing 7.4, add the definition of **bazquux** using the def→| trigger, then add the final line shown by using autocomplete to type **FooBar** and **bazquux**. (Type the interstitial **.new.** by hand.)

3. Using tab triggers and autocomplete, make a file called **greeter.rb** with the contents shown in Listing 7.5.

4. By cutting and pasting the text for the **hello** definition and indenting the block, transform Listing 7.5 into Listing 7.6.

Listing 7.3: Creating a class using a tab trigger.
~/foo.rb

```
class FooBar

end
```

Listing 7.4: Using autocomplete to make a class name.
~/foo.rb

```
class FooBar
  def bazquux
    puts "Baz quux!"
  end
end

FooBar.new.bazquux
```

Listing 7.5: A proto-Greeter class in Ruby.
~/greeter.rb

```
class Greeter
end

def hello(location)
  puts "hello, #{location}!"
end

Greeter.new.hello("world")
```

Listing 7.6: A completed Greeter class in Ruby.

```
class Greeter
  def hello(location)
    puts "hello, #{location}!"
  end
end

Greeter.new.hello("world")
```

7.3 Writing an Executable Script

As a practical application of the material in Section 7.2, in this section we're going to write something that's actually useful: a *shell script* designed to kill a program as safely as possible. (A *script* is a program that is typically used to automate common tasks, but the detailed definition isn't important at this stage.) En route, we'll cover the steps needed to add this script to our command-line shell.

As discussed in Box 3.2, Unix user and system tasks take place within a well-defined container called a *process*. Sometimes, one of these processes will get stuck or otherwise misbehave, in which case we might need to terminate it with the **kill** command, which sends a *terminate code* to kill the process with a given id:

```
$ kill -15 12241
```

(See the discussion in Box 3.2 for more on how to find this id on your system.) Here we've used the terminate code **15**, which attempts to kill the process as gently as possible (meaning it gives the process a chance to clean up any temporary files, complete any necessary operations, etc.). Sometimes terminate code **15** isn't enough, though, and we need to escalate the level of urgency until the process is well and truly dead. It turns out that a good sequence of codes is **15**, **2**, **1**, and **9**. Our task is to write a command to implement this sequence, which we'll call **ekill** (for "escalating kill"), so that we can kill a process as shown in Listing 7.7.

Listing 7.7: An example of using **ekill** (to be defined).

```
$ ekill 12241
```

As with the Ruby example in Section 7.2, don't worry about the details of the code; focus instead on the mechanics of the text editing.

As preparation for adding **ekill** to our system, we'll first make a new directory in our home directory called **bin** (for "binary"):

```
$ mkdir ~/bin
```

(It's possible that this directory already exists on your system, in which case you'll get a harmless warning message.) We'll then change to the **bin** directory and open a new file called **ekill**:

```
$ cd ~/bin
$ atom ekill
```

The **ekill** script itself starts with a "shebang" line (pronounced "shuh-BANG", from "shell" and "bang", with the latter being the common pronunciation of the exclamation point **!** (Box 3.1)):

```
#!/bin/bash
```

This line tells our system to use the shell program located in **/bin/bash** to execute the script. The **bash** program corresponds to the Bourne-again shell (Bash) mentioned in Section 5.3, and in this context a shell script is often called a *Bash script*.[8] Despite appearances, here the hash symbol **#** is *not* a comment character, which is potentially confusing because (as in Ruby) **#** is the character ordinarily used for a Bash comment line. Indeed, the initial version of our script includes several comment lines, as shown in Listing 7.8.

Listing 7.8: A custom escalating kill script.
~/bin/ekill

```
1  #!/bin/bash
2
3  # Kill a process as safely as possible.
4  # Tries to kill a process using a series of signals with escalating urgency.
5  # usage: ekill <pid>
6
7  # Assign the process id to the first argument.
8  pid=$1
9  kill -15 $pid || kill -2 $pid || kill -1 $pid || kill -9 $pid
```

Apart from the shebang in line 1, all other uses of **#** introduce comments. Then, line 8 assigns the process id **pid** to **$1**, which in a shell script is the first argument to the command, e.g., **12241** in Listing 7.7. Line 9 then uses the "or" operator **||** to execute the **kill** command using the code **15** or **2** or **1** or **9**, stopping on the first successful **kill**. (Again, don't worry if you find this confusing; I include the explanation for completeness, but at this stage there's no need to understand the details.)

After typing the contents of Listing 7.8 into the script file, one thing you might notice is that the result has no syntax highlighting, as seen in Figure 7.18. This is because, unlike **README.md** (Section 6.2) and **hello.rb** (Section 7.2), the name

---

8. To learn how to write this same script using Zsh, see "Using Z Shell on Macs with the Learn Enough Tutorials" (https://news.learnenough.com/macos-bash-zshell).

ekill has no filename extension. Although some people would use a name like **ekill.sh** for shell scripts like this one—which would in fact allow our editor to highlight the syntax automatically—using an explicit extension on a shell script is a bad practice because the script's name is the user interface to the program. As users of the system, we don't care if **ekill** is written in Bash or Ruby or C, so calling it **ekill.sh** unnecessarily exposes the implementation language to the end-user. Indeed, if we wrote the first implementation in Bash but then decided to rewrite it in Ruby and then in C, every program (and programmer) using the script would have to change the name from **ekill.sh** to **ekill.rb** to **ekill.c**—an annoying and avoidable complication.

Even though we've elected not to use a filename extension for the **ekill** script, we'd still like to get syntax highlighting to work. One way is to click on "Plain Text" in the lower right-hand corner of the editor (Figure 7.18) and change the highlighting language to the one we're using. This requires us to *know* the language, though, and it would be nicer if we could get the editor to figure it out automatically. Happily, we can arrange exactly that, simply by closing the file and opening it again. To do this, click on the X to close the **ekill** tab (or press ⌘W) and then re-open it from the command line:

```
$ atom ekill
```

Because of the shebang line in Listing 7.8, Atom infers that the file is a Bash script. As a result, the detected file type changes from "Plain Text" to "Shell Script", and syntax highlighting is activated (Figure 7.19).

At this point, we have a complete shell script, but typing **ekill <pid>** at the command line still won't work. To add **ekill** to our system, we need to do two things:

1. Make sure the **~/bin** directory is on the system *path*, which is the set of directories where the shell program searches for *executable* scripts.
2. Make the script itself executable.

The list of directories on the path can be accessed via the special **$PATH** variable at the command line:

```
$ echo $PATH
```

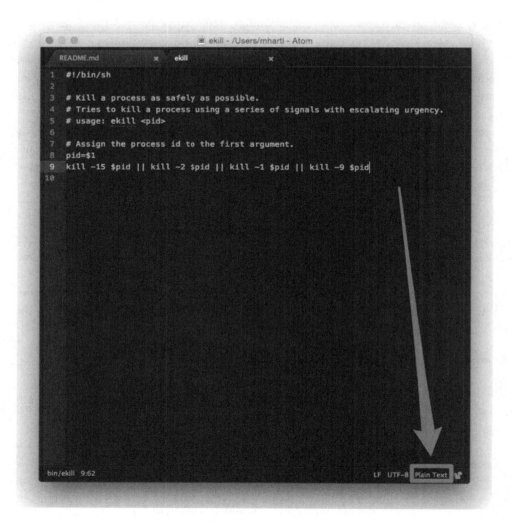

Figure 7.18: The `ekill` script with no syntax highlighting.

If **~/bin** is on the list, you can skip this step, but it does no harm to follow it.

Note: The literal directory **~/bin** won't appear in the **$PATH** list; instead the tilde will be expanded to your particular home directory. For me, **~/bin** is the same as **/Users/mhartl/bin**, so that's what appears in my **PATH**, but it will be different for you.

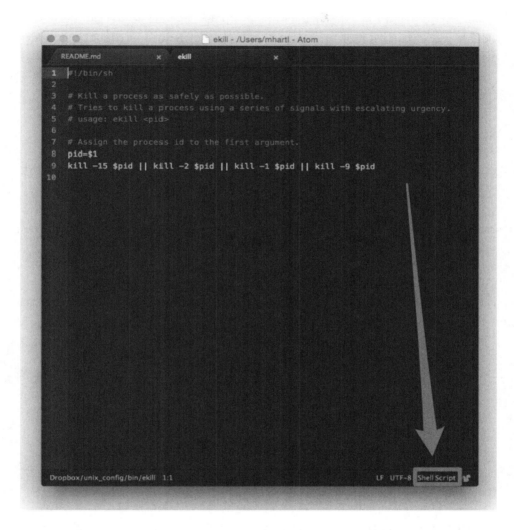

Figure 7.19: The `ekill` script with syntax highlighting and a new detected file type.

To make sure **~/bin** is on the path, we'll edit the Bash profile file, which is related to the **.bashrc** file we saw in Section 5.3. Open **~/.bash_profile** as follows:

```
$ atom ~/.bash_profile
```

Then add the **export** line shown in Listing 7.9. If the **source** line isn't already present, you should add that as well—it ensures that any aliases defined in `.bashrc` are added when `.bash_profile` gets run.[9]

Listing 7.9: Adding ~/**bin** to the path.
~/.bash_profile

```
export PATH="~/bin:$PATH"
source ~/.bashrc
```

This uses the Bash **export** command to add **~/bin** to the current path. (It's worth noting that some systems use the *environment variable* **$HOME** in place of **~**, but the two are synonyms. If for any reason **~** doesn't work for you, it's worth trying **$HOME** instead, as in **$HOME/bin:$PATH**.)

To use it, we need to use **source** as in Section 5.4:

```
$ source ~/.bash_profile
```

To make the resulting script executable, we need to use the "change mode" command **chmod** to add the "execute bit" **x** as follows:

```
$ chmod +x ~/bin/ekill
```

At this point, we can verify that the **ekill** script is ready to go using the **which** command (Section 3.1):

```
$ which ekill
```

The result should be the full path to **ekill**, which on my system looks like this:

```
$ which ekill
/Users/mhartl/bin/ekill
```

---

9. Keeping track of when each of `.bashrc` and `.bash_profile` gets run can get complicated, and in many cases editing either file works.

On some systems, running **source** on **.bash_profile** might not be sufficient to put **ekill** on the path, so if **which ekill** returns no result then you should try exiting and restarting the shell program to reload the settings.

As you can see by typing **ekill** by itself at the command line, the current behavior is confusing if we neglect to include a process id:

```
$ ekill
<confusing error message>
```

To make **ekill** friendlier in this case, we'll arrange to print a usage message to the screen if the user neglects to include a process id. We can do this with the code in Listing 7.10, which I recommend typing in rather than copying and pasting. When writing the **if** statement, I especially recommend trying `if →|` to see if your editor comes with a tab trigger for making Bash **if** statements.

Listing 7.10: An enhanced version of the escalating kill script.
~/bin/ekill

```
#!/bin/bash

# Kill a process as safely as possible.
# Tries to kill a process using a series of signals with escalating urgency.
# usage: ekill <pid>
# If the number of arguments is less than 1, exit with a usage statement.
if [[ $# -lt 1 ]] ; then
    echo "usage: ekill <pid>"
    exit 1
fi
# Assign the process id to the first argument.
pid=$1
kill -15 $pid || kill -2 $pid || kill -1 $pid || kill -9 $pid
```

After adding the code in Listing 7.10, running **ekill** without an argument should produce a helpful message:

```
$ ekill
usage: ekill <pid>
```

All we have left to do is to verify that **ekill** can actually be used to kill a process. This is left as an exercise (Section 7.3.1.)

7.3.1   Exercises

1. Let's test the functionality of **ekill** by making a process that hangs and applying the lessons from grepping processes (Box 3.2). We'll start by opening two terminal tabs. In one tab, type **tail** to get a process that just hangs. In the other tab, use **ps aux | grep tail** to find the process id, then run **ekill <pid>** (substituting the actual id for **<pid>**). In the tab running **tail**, you should get something like "Terminated: 15" (Figure 7.20).

2. Write an executable script called **hello** that takes in an argument and prints out "Hello" followed by the argument. Be sure to **chmod** the script so it can run properly. *Hint*: Use the **echo** command. *Bigger hint*: Bash scripts *interpolate*

Figure 7.20: The result of using **ekill** to kill a **tail** process.

dollar-sign variables into strings, so the **$1** variable from Listing 7.8 can be used in a string like this: **"Hello, $1"**.

7.4 Editing Projects

So far we've used our text editor to edit single files, but it can also be used to edit entire projects all at once. As an example of such a project, we'll download the sample application from the 3rd edition of the *Ruby on Rails Tutorial* (https://www.railstutorial.org/book). We won't be running this application, but it will give us a large project to work with. As in Section 5.6 and Section 6.2, we'll use the **curl** command to download the file to our local disk:

```
$ cd
$ curl -OL https://source.railstutorial.org/sample_app.zip
```

As indicated by the **.zip** filename extension, this is a ZIP file, so we'll unzip it (using the **unzip** command) and then **cd** into the sample app directory:

```
$ unzip sample_app.zip
   creating: sample_app_3rd_edition-master/
   .
   .
   .
$ cd sample_app_3rd_edition-master/
```

The way to open a project is to use a text editor to open the entire directory. Recall from Section 4.3 that **.** ("dot") is the current directory, which means that we can open it using "atom dot":

```
$ atom .
```

The resulting text editor window includes the directory structure for our project, called a "tree view", as seen in Figure 7.21. We can toggle its display using the View menu or a keyboard shortcut (Figure 7.22).

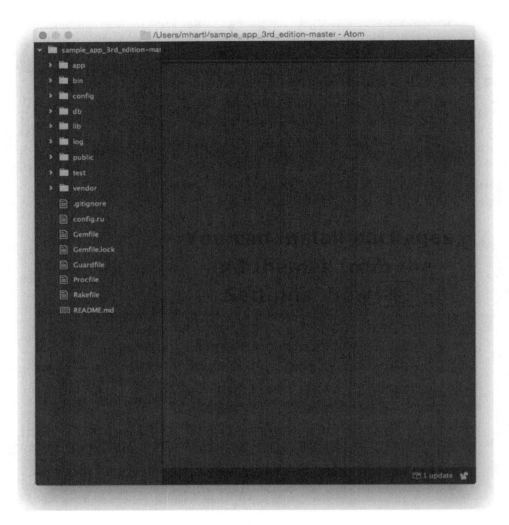

Figure 7.21: The Rails Tutorial sample app in Atom.

7.4.1 Fuzzy Opening

It's possible to open a file by double-clicking on it in the tree view, but in a project with a lot of files this is often cumbersome, especially when the file is buried several subdirectories deep. A convenient alternative is *fuzzy opening*, which lets us open files

Figure 7.22: Toggling the tree view.

by hitting (in Atom) ⌘P and then typing some of the letters in the filename we want. For example, we can open a file called **users_controller_test.rb** by typing, say, "userscon" and then selecting from the drop-down menu, as shown in Figure 7.23. The letters don't have to be contiguous in the filename, though, so typing "uctt" (for **u**sers **c**ontroller **t**est) will also work, as seen in Figure 7.24.

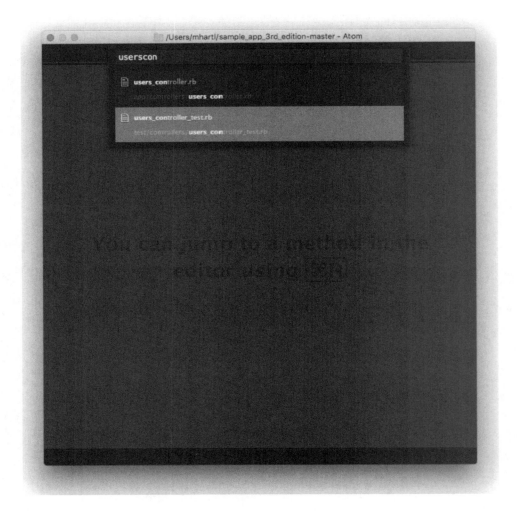

Figure 7.23: One way to open a file with fuzzy opening.

As a result of opening multiple files in a project, you will generally have multiple tabs open in your editor (Figure 7.25). I recommend learning the keyboard shortcuts to switch between them, which are typically things like ⌘1, ⌘2, etc. (By the way, this trick also works in many browsers, such as Chrome and Firefox.)

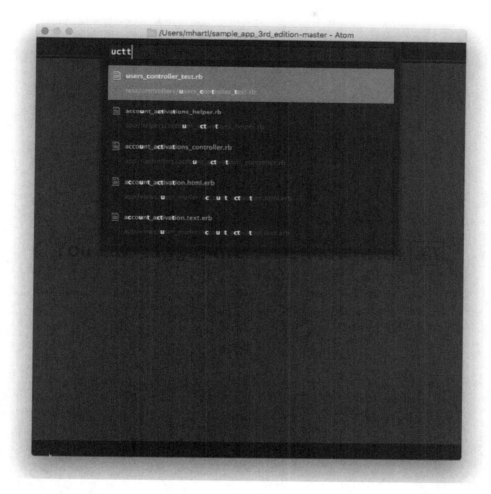

Figure 7.24: A second way to open a file with fuzzy opening.

7.4.2   Multiple Panes

The default editor view we've seen in most of the previous examples consists of a single *pane* (as in "window pane"), but it's often convenient to split the editor into multiple panes so that we can see more than one file at a time (Figure 7.26). I especially like to use different panes for different types of files, such as using the left pane for test code

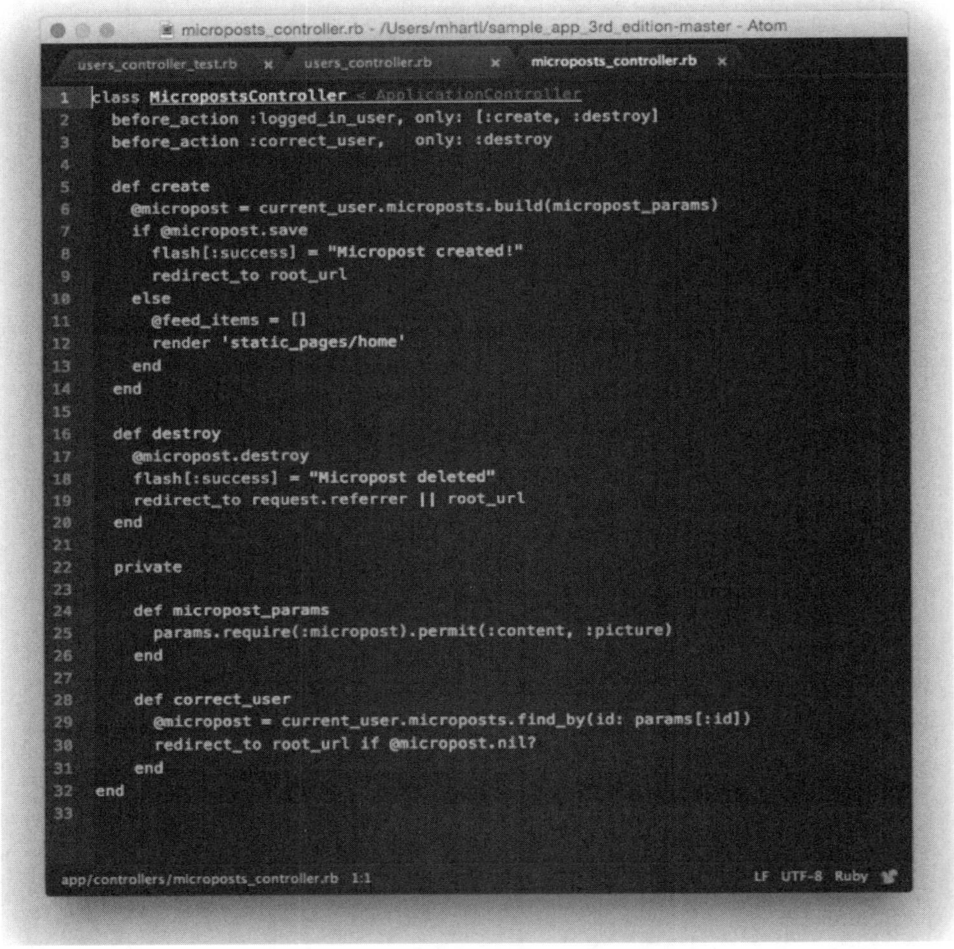

Figure 7.25: Opening multiple tabs.

and the right pane for application code. It's also often useful to open the same file in two different panes (Figure 7.27); as I wrote in the meta–tutorial *Learn Enough Tutorial Writing to Be Dangerous* (https://www.learnenough.com/tutorial-writing-tutorial):

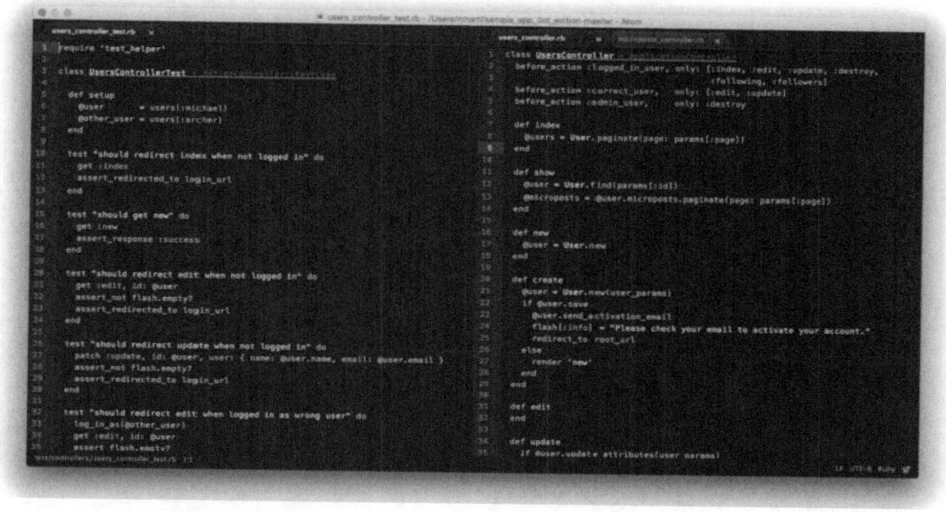

Figure 7.26: Using multiple panes.

When searching through the document for whatever reason (to fix an error, look up a label for a cross-reference, find a particular string, etc.), it's usually inconvenient to move the cursor and hence lose our place. In this context, it's useful to have the same file open in two different text editor windows… This way, we can use one pane as the main writing area and the other pane as a sort of "random access" window for moving around in the document.

Note: "Panes" are sometimes called "Groups" (e.g., in Sublime Text).

7.4.3   Global Find and Replace

We saw in Section 6.8 how to find and optionally replace content in a single file. When editing projects, it's often useful to be able to do a *global* find and replace across multiple files. As usual, most editors have both a menu item (Figure 7.28) and a keyboard shortcut (often ⇧⌘F).

An example of global find appears in Figure 7.29, which searches for the string "@user" in all project files. The command to globally replace this with "@person" appears in Figure 7.30.

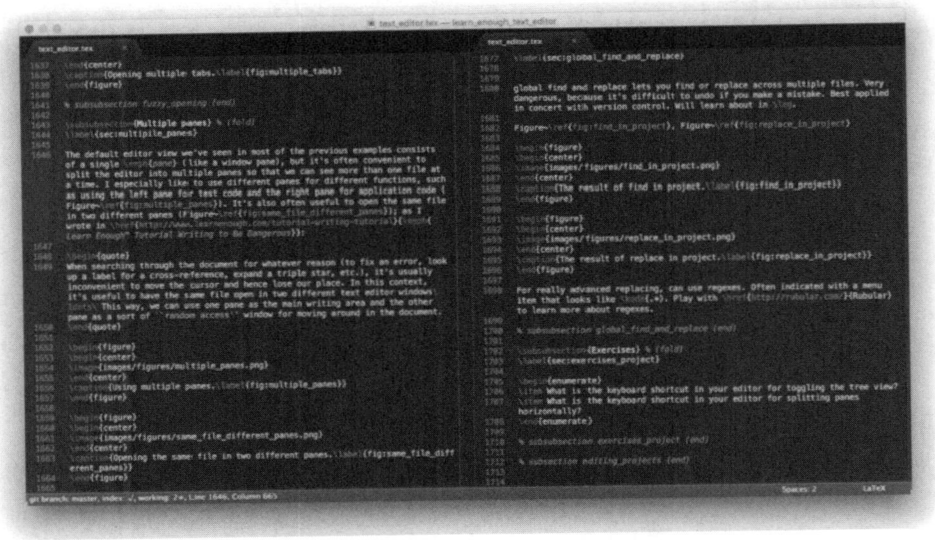

Figure 7.27: Opening the same file in two different panes.

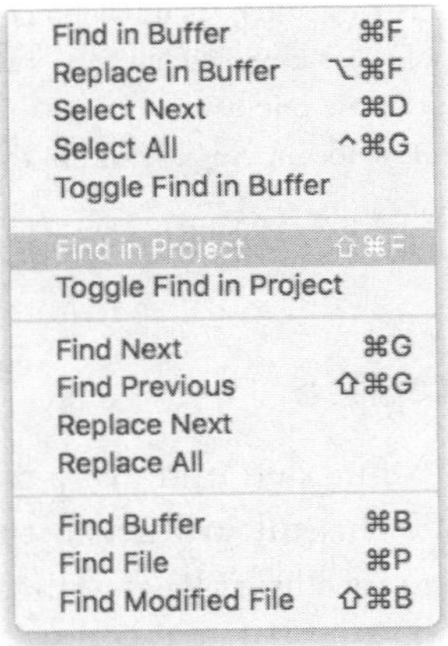

Figure 7.28: A menu item for global find and replace.

Figure 7.29: The result of finding in project.

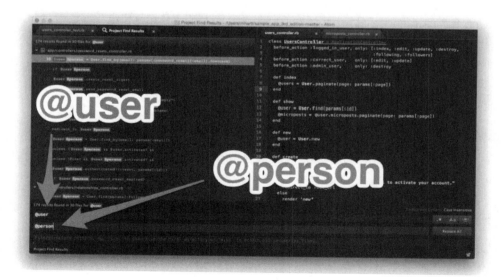

Figure 7.30: The result of replacing in project.

For really advanced replacing, we can use a mini-language for text pattern matching called *regular expressions* (or *regexes* for short), as mentioned briefly in Section 3.4. Let's see how to use regexes to add an annotation to all function definitions in the project, changing

```
def foo
```

to

```
def foo    # function definition
```

and

```
def bar
```

to

```
def bar    # function definition
```

My favorite way to build up regular expressions is using a web application like regex101 (as mentioned briefly in Section 3.4), which lets us create regexes interactively (Figure 7.31). Moreover, such resources typically include a quick reference to assist us in finding the code for matching particular patterns (Figure 7.32).

We can use the reference in Figure 7.32 to discover a regex for **def** followed by any sequence of characters, which looks like this:

```
def .*
```

Here **.** represents "any character", while ***** matches zero or more of them. Doing a global search using the regex **def** **.*** matches all the function definitions in the project, as seen in Figure 7.33. Note that in most editors you'll have to enable regex matching by clicking the regex match icon (.* in Figure 7.33).

We can do the replacement mentioned above using parentheses to create two *match groups*:

```
(def) (.*)
```

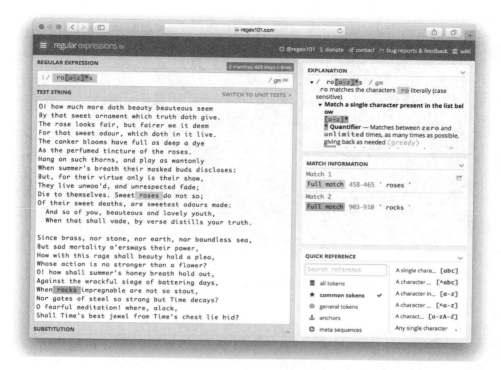

Figure 7.31: An online regex tester.

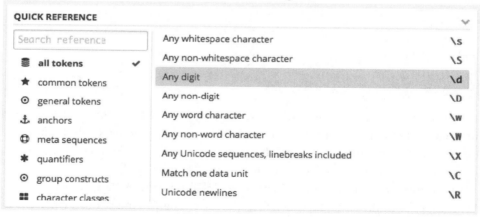

Figure 7.32: A close-up of the regex reference.

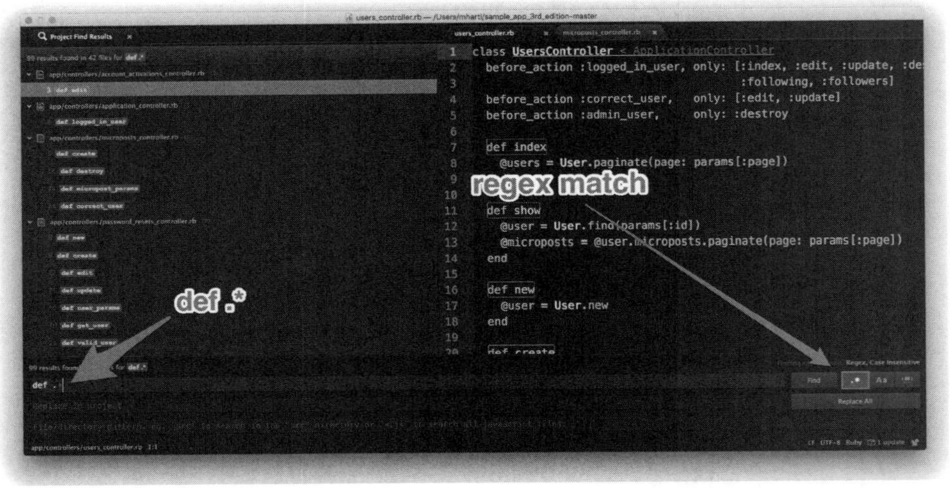

Figure 7.33: Matching a regular expression.

The first match group here is just the constant string **def**, while the second is whatever the function definition happens to be. (These match groups also appear in Figure 7.31.) Inside the "Replace" field in the editor, we can reference these groups using special dollar-sign match numbers, so that we can replace

```
(def) (.*)
```

with

```
$1 $2    # function definition
```

For example, when matching **def foo**, **$1** is **def** and **$2** is **foo**; when matching **def bar**, **$1** is still **def**, but **$2** is **bar**. This means we can annotate all the function definitions at the same time using the commands shown in Figure 7.34. Actually completing this replacement is left as an exercise (Section 7.4.4).

One thing to bear in mind when using global find and replace is that it can be hard to undo. In the case of a single file, it's easy enough to undo a bad replacement

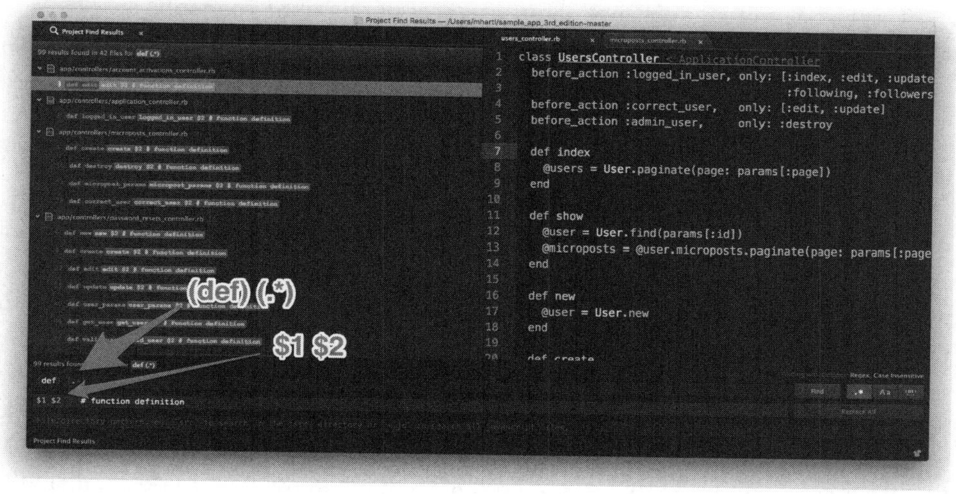

Figure 7.34: Using a match grouping.

with ⌘Z (Section 6.6), but when replacing across multiple files we have to run ⌘Z in *every* affected file, which could be dozens. As a result, I recommend using global find and replace with great caution, and preferably in combination with a version control system such as Git. My general practice is to make a *commit* before any global search and replace so that I can easily undo it if there turns out to be a mistake. (See *Learn Enough Git to Be Dangerous* (https://www.learnenough.com/git) for more information.)

7.4.4 Exercises

1. What is the keyboard shortcut in your editor for toggling the tree view?
2. What is the keyboard shortcut in your editor for splitting panes horizontally?
3. In the Rails Tutorial sample app project, open the file **static_pages_controller.rb** using fuzzy opening.
4. Use global find to find all occurrences of the string **@user**.

5. Use global replace to change all occurrences of **@user** to **@person**.

6. Use a regex match to annotate all function definitions with **# function definition** as described in the text.

7.5 Customization

All good text editors are highly customizable, but the options are highly editor-dependent. The most important things are (a) to know what kind of customization is possible and (b) to apply your technical sophistication (Box 5.2) to figure out how to make the desired changes.

For example, one student of the *Ruby on Rails Tutorial* wrote in asking about the dark background in the Cloud9 editor (e.g., Figure 6.5), wondering if it was possible to use a light background instead. I responded that it was almost certainly possible to change to a light background, even though I didn't know how to do it offhand. I knew that every good programmer's editor has multiple highlighting color schemes, font sizes, tab sizes, etc., so I was confident I could figure out how to change the background color on the Cloud9 editor. And indeed, by clicking around and looking for promising menu items (a textbook application of Box 5.2), I was able to discover the answer (Preferences > Themes > Syntax Theme > Cloud9 Day), as shown in Figure 7.35.

Another feature common to good text editors is some sort of package system. For example, we saw in Section 6.2.2 that Atom comes with a built-in package to preview Markdown, but in Sublime Text we need to install a separate package called Package Control (https://packagecontrol.io/) to do it. One way to find new packages is to Google around for more information, leading to a site like that shown in Figure 7.36. The result is a new option, Sublime Text > Preferences > Package Control, as shown in Figure 7.37 and Figure 7.38.

Most editors allow you to create your own packages of commands, as well as often supporting *snippets* that let you define your own tab triggers (Section 7.1). These are advanced topics, so I recommend deferring them for now. Once you start becoming annoyed by having to repeatedly type the same boilerplate (as in, e.g., Listing 7.11), Google around to figure out how to add custom commands to your editor. (The code in Listing 7.11 is generated using the custom Sublime Text tab trigger **clist** (for "code listing"), which I have also ported to Atom.)

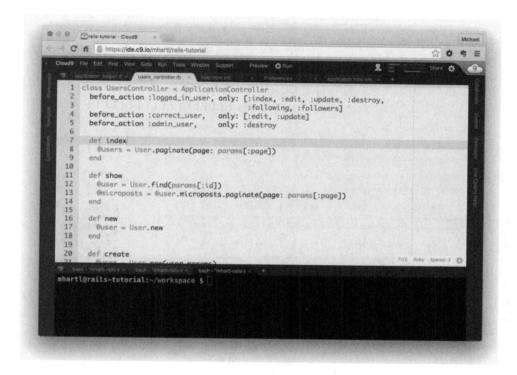

Figure 7.35: The Cloud9 editor with a light background.

Listing 7.11: The boilerplate for a code listing in this document.

```
\begin{codelisting}
\label{code:}
\codecaption{}
%= lang:
\begin{code}

\end{code}
\end{codelisting}
```

7.5.1   Exercises

1. Figure out how to change the syntax highlighting theme in your editor. Use the file from Listing 7.6 to confirm the change.

Figure 7.36: Searching for a Sublime Text package.

Figure 7.37: Sublime Text's Package Control menu item.

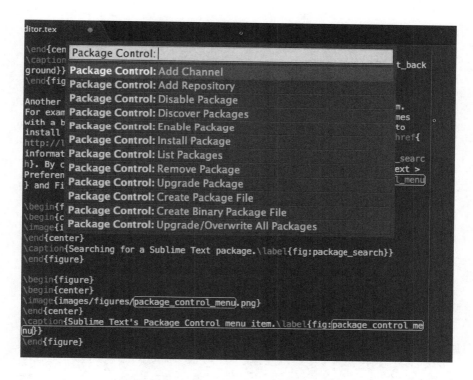

Figure 7.38: Sublime Text's Package Control.

2. In Atom, figure out how to install the `minimap` package. What does this package do? The result for **sonnets.txt** should look something like Figure 7.39.

7.6 Summary

- Autocomplete and tab triggers make it easy to type lots of text quickly.

- All good text editors have special features to support writing computer source code, including syntax highlighting, commenting out, indenting & dedenting, and goto line number.

- Many programmers think it's perfectly fine to have lines that are more than 80 columns across, but they are wrong. (Speaking of which, there's nothing quite so entertaining as a holy war (Box 5.4)…)

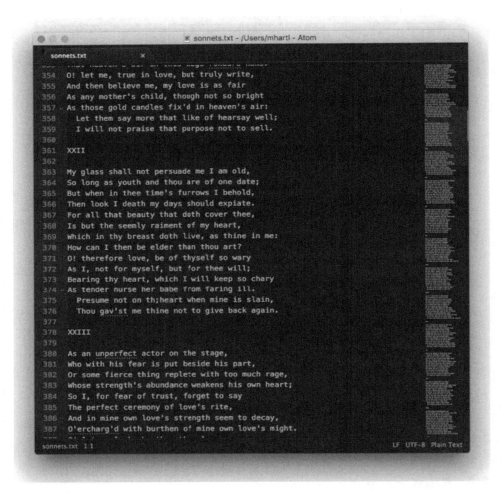

Figure 7.39: The result of adding `minimap` to Atom.

- Once you know how to use the command line and a text editor, it's easy to add custom shell scripts to your system.

- It's common to open entire projects (such as Ruby on Rails applications) all at once using the command line.

- Fuzzy opening is useful when editing projects with large numbers of files.

- Using multiple panes allows the editor to display more than one file at a time.
- Global find and replace is dangerous but powerful.
- All good programmer's editors are extensible and customizable.

Important commands from this chapter are summarized in Table 7.1.

Table 7.1: Important commands from Chapter 7.

Command	Description
Select + ⌘/	Toggle commenting out
Select + →|	Indent
Select + ⇧ →|	Dedent
^G	Goto line number
⌘W	Close a tab
$ echo $PATH	Show the current path variable
$ chmod +x <filename>	Make `filename` executable
$ unzip <filename>.zip	Unzip a ZIP archive
⌘P	Fuzzy opening
⌘1	Switch focus to tab #1
⇧⌘F	Global find and replace

7.7 Conclusion

Congratulations! You now know enough text editor to be *dangerous*. If you continue down this technical path, you'll keep getting better at using text editors for years to come, but with the material in this tutorial you've got a great start. For now, you're probably best off working with what you've got, applying your technical sophistication (Box 5.2) when necessary. Once you've got a little more experience under your belt, I recommend seeking out resources specific to your editor of choice. To get you started, here are some links to documentation for the editors mentioned in this tutorial:

- Atom docs (https://atom.io/docs)
- Sublime Text docs (https://www.sublimetext.com/docs/)
- Cloud9 editor docs (https://aws.amazon.com/cloud9/details/)

As a reminder, *Learn Enough Text Editor to Be Dangerous* is just one in a series of tutorials (https://www.learnenough.com/courses) designed to teach the fundamentals of software development. The next step in the series is Part III of *Learn Enough Developer Tools to Be Dangerous*, and the full sequence appears as follows:

1. ***Learn Enough Developer Tools to Be Dangerous***
   (a) Part I: *Learn Enough Command Line to Be Dangerous*
   (b) Part II: *Learn Enough Text Editor to Be Dangerous* (you are here)
   (c) Part III: *Learn Enough Git to Be Dangerous*
2. **Web Basics**
   (a) *Learn Enough HTML to Be Dangerous*
   (b) *Learn Enough CSS & Layout to Be Dangerous* (https://www.learnenough.com/css-and-layout)
   (c) *Learn Enough JavaScript to Be Dangerous* (https://www.learnenough.com/javascript)
3. **Application Development**
   (a) *Learn Enough Ruby to Be Dangerous*
   (b) *Ruby on Rails Tutorial*
   (c) *Learn Enough Action Cable to Be Dangerous* (https://www.learnenough.com/action-cable) (optional)

Good luck!

PART III

Git

CHAPTER **8**

Getting Started with Git

Learn Enough Git to Be Dangerous is Part III of *Learn Enough Developer Tools to Be Dangerous*. So far in this tutorial, we have covered two skills essential for software developers and those who work with them: In Part I, *Learn Enough Command Line to Be Dangerous*, we learned how to use the Unix command line, and in Part II, *Learn Enough Text Editor to Be Dangerous*, we learned how to use text editors. Part III covers a third essential skill: *version control*.

As with its two predecessors, *Learn Enough Git to Be Dangerous* doesn't even assume you're familiar with the *category* of application, so if you're unsure about what "version control" is, you're in the right place. Even if you are already familiar with the subject, it's likely you'll still learn a lot from this tutorial. Either way, learning the material in Part III of *Learn Enough Developer Tools to Be Dangerous* prepares you for the other Learn Enough tutorials (https://www.learnenough.com/) while enabling an astonishing variety of applications—including a special surprise bonus at the end (Box 8.1).

Box 8.1: Real Artists Ship

As legendary Apple cofounder Steve Jobs once said: *Real artists ship.* What he meant was that, as tempting as it is to privately polish in perpetuity, makers must *ship* their work—that is, actually finish it and get it out into the world. This can be scary, because shipping means exposing your work not only to fans but also to critics. "What if people don't like what I've made?" *Real artists ship.*

It's important to understand that shipping is a separate skill from making. Many makers get good at making things but never learn to ship. To keep this from happening to us, starting in *Learn Enough Git to Be Dangerous* we're going to ship at least one thing in every Learn Enough tutorial. In fact, in this tutorial we'll actually ship *two* things—a public *Git repository* and a surprise bonus that will give you bragging rights with all of your friends.

Version control solves a problem that might look familiar if you've ever seen Word documents or Excel spreadsheets with names like `Report_2014_1.doc`, `Report_2014_2.doc`, `Report_2014_3.doc`, or `budget-v7.xls`. These cumbersome names indicate how annoying it can be to track different versions of documents. Nowadays, applications like Word do sometimes offer built-in version tracking, but such features are tightly coupled to the underlying application, and aren't useful for any other document types. Many technical applications (including most websites and programming projects) require a general solution to the problem of versions.

A version control system, or *VCS*, provides an automatic way to track changes in software projects, giving creators the power to view previous versions of files and directories, develop speculative features without disrupting the main development, securely back up the project and its history, and collaborate easily and conveniently with others. In addition, using version control also makes deploying production websites and web applications much easier. As a result, fluency in at least one version control system is an essential component of *technical sophistication* (Box 8.2), a useful skill for developer, designer, and manager alike. This applies especially to the version control system covered in this tutorial, called *Git*.

Box 8.2: Technical Sophistication

A principal theme of the Learn Enough tutorials is the development of *technical sophistication*, the combination of hard and soft skills that make it seem like you can magically solve any technical problem (as illustrated in "Tech Support Cheat Sheet" (https://m.xkcd.com/627/) from xkcd). *Learn Enough Git to Be Dangerous* is important for developing these skills because being able to use at least one modern version control system is an essential component of technical sophistication.

In the context of Git, technical sophistication includes several things. Many Git commands print various details to the terminal screen; technical sophistication lets you figure out which ones to pay attention to and which to ignore. There are also many Git-related resources on the Web, which among other things means that Google searches are often useful for figuring out the exact command you need at a particular time. Technical sophistication lets you figure out the best search terms for finding the answer you're looking for; e.g., if you need to delete a remote branch (Section 11.3.1), Googling for "git delete remote branch" is a good bet to turn up something useful. Finally, repository hosting sites like GitHub, GitLab, and Bitbucket typically include commands to help guide you through various setup tasks, and technical sophistication gives you the confidence to follow the steps even if you don't understand every detail.

One helpful command for learning Git is `git help`, which by itself gives general guidelines on Git usage, and when applied to a specific command gives further information on that command. For example, `git help add` shows details about the `git add` command. The output of `git help` is similar to the man pages covered in Section 1.4: full of useful but often obscure information. As always, use your technical sophistication to help make sense of it.

Version control has evolved considerably over the years. The family line leading to Git includes programs called RCS, CVS, and Subversion, and there are many current alternatives as well, including Perforce, Bazaar, and Mercurial. I mention these examples not because you need to know what they are, but only to show what a bewildering variety there is. What's worse, when you choose a version control system, you really *commit* to it,[1] and it is often difficult to switch from one to another. Happily, in the last few years an undisputed winner has emerged in the open-source VCS wars: Git. This victory is the main reason this tutorial is called *Learn Enough Git to Be Dangerous* rather than *Learn Enough Version Control to Be Dangerous*. Nevertheless, many of the ideas here are quite general, and if by some chance you need to use a different VCS, this tutorial will still provide a useful introduction to the subject.

Originally developed by Linux creator Linus Torvalds[2] to host the Linux kernel, Git is a command-line program that is designed in the Unix tradition (which is why a familiarity with the Unix command line is an important prerequisite). Git has a combination of power, speed, and community adoption that leave it few rivals, but it can be tricky to learn, and other Git tutorials have a tendency to introduce lots of

1. Pun intended. If you don't get it, don't worry—by the end of this tutorial, you will.

2. *Git* is a mildly insulting British slang term for a stupid or annoying person, and Linus likes to joke that he named both Linux and Git after himself.

heavy theory, which can be interesting to learn but in practice is really only understood by a tiny handful of Git users (as illustrated in "Git" (https://m.xkcd.com/1597/) via the webcomic xkcd). The good news is that the set of Git commands needed to be productive is relatively small; there are some pointers to more advanced and theory-oriented resources listed in Section 11.7, but in this tutorial we focus on the essential commands needed to be *dangerous*.

Note: If you're using macOS, you should follow the instructions in Box 2.3 at this time.

8.1 Installation and Setup

The most common way to use Git is via a command-line program called **git**, which lets us transform an ordinary Unix directory into a *repository* (or *repo* for short) that enables us to track changes to our project.[3] In this section, we'll begin by installing Git (if necessary) and doing some one-time configuration.

Before doing anything else, we first need to check to see if Git is installed on the current system. As a reminder, we're working in the Unix tradition, so it is strongly recommended that you use macOS or Linux. Microsoft Windows users are encouraged to set up a Linux-compatible development environment by following the steps in Section A.3.3 or by using the Linux-based cloud IDE discussed in Section A.2.

The easiest way to check for Git is to start a terminal window and use **which** (Section 3.1) at the command line to see if the **git** executable is already present:

```
$ which git
/usr/local/bin/git
```

If the result is empty or if it says the command is not found, it means you have to install Git manually. To do this, follow the instructions at "Getting Started – Installing Git" (https://git-scm.com/book/en/v2/Getting-Started-Installing-Git) in the official Git documentation. (This will likely give you an opportunity to apply some technical sophistication (Box 8.2).)

---

3. As we'll see in Section 8.2, Git uses a special hidden directory called **.git** to track changes, but at the level of this tutorial these details aren't important.

The next step is to ensure that you have a sufficiently recent version of Git, which you can check as follows:

```
$ git --version
git version 2.31.1     # should be at least 2.28.0
```

If the version isn't recent enough, then you should either install Git via the instructions at "Getting Started – Installing Git" or follow one of the suggestions below:

- Cloud IDE: If using the cloud IDE recommended in Section A.2, run the command in Listing 8.1.

- macOS: If using macOS, run the command in Listing 8.2. (If you don't have Homebrew installed, first follow the instructions to install Homebrew (https://brew.sh/).)

Listing 8.1: Upgrading Git on the cloud IDE.

```
$ source <(curl -sL https://cdn.learnenough.com/upgrade_git)
```

Listing 8.2: Upgrading Git on macOS.

```
$ brew upgrade git
```

After installing Git but before starting a project, we need to perform a few one-time setup steps, as shown in Listing 8.3. These are *global* setups, meaning you only have to do them once per computer. (Don't worry about the meaning or structure of these commands at this stage.)

Listing 8.3: One-time global configuration settings.

```
$ git config --global user.name "Your Name"
$ git config --global user.email your.email@example.com
$ git config --global init.defaultBranch main
```

The first two configuration settings allow Git to identify changes in your projects by name and email address, which is especially helpful when collaborating with others (Chapter 11). Note that the name and email you use in Listing 8.3 will be viewable in any projects you make public, so don't expose any information you'd rather keep private.

The third line in Listing 8.3 sets the default Git branch name to **main**, which is the current recommended default. You should be aware that the default branch name for the first 15+ years of Git's existence was **master**, so you will invariably encounter many Git repositories that use **master** instead of **main**.[4] (Being able to deal with this sort of situation is a hallmark of technical sophistication (Box 8.2).) See the Learn Enough blog post "Default Git Branch Name with Learn Enough and the Rails Tutorial" (https://news.learnenough.com/default-git-branch-name-with -learn-enough-and-the-rails-tutorial) for more information.

In addition to the configuration in Listing 8.3, *Learn Enough Git to Be Dangerous* includes some optional advanced setup (Section 11.6) that I recommend you complete at some point. If you already have some familiarity with Git, or if you're an experienced user of the Unix command line, I recommend completing the steps in Section 11.6 at this time, but otherwise I recommend deferring the advanced setup until later.

8.1.1 Exercises

1. Run **git help** at the command line. What is the first command listed?

2. There's a chance that the full output of **git help** was too big to fit in your terminal, with most of it just scrolling by. What's the command to let us navigate the output of **git help** interactively? (On some systems, you can use the mouse to scroll back in the terminal window, but it's unwise to rely on this fact.) *Hint*: Pipe (Section 3.2.1) the output to **less** (Section 3.3).

3. Git stores global configuration settings in a hidden text file located in your home directory. By inspecting the file **~/.gitconfig** with a tool of your choice (**cat**, **less**, a text editor, etc.), confirm that the configuration set up by Listing 8.3 corresponds to simple text entries in this file.

---

4. The screencasts that accompany this book use **master** since they were made at a time when the use of that default branch name was universal, but you can safely substitute **main** for **master** in the videos as well.

8.2 Initializing the Repo

Now it's time to start creating a project and put it under version control with Git. To see how Git works and what benefits it brings, it helps to have a concrete application in mind, so we'll be tracking changes in a simple project consisting of a small website consisting of two pages, a Home page and an About page.[5] We'll begin by making a directory with the generic name **website** inside a repositories directory called **repos**:

```
[~]$ mkdir -p repos/website
```

Here we've used the "make directory" command **mkdir** (Section 4.2) together with the **-p** option, which arranges for **mkdir** to create intermediate directories as required (in this case, **repos**). Note also that I've included the current directory in the prompt (in this case, **[~]**) as arranged by the configuration in Listing 11.15.

After making the directory, we can **cd** into it as follows:

```
[~]$ cd repos/website/
[website]$
```

(Recall from Box 2.4 that you can use tab completion when changing directories, so in real life I would probably type something like **cd re→w→.**)

Even though the **website** directory is empty, we can already convert it to a *repository*, which you can think of as a sort of enhanced Unix directory with the additional ability to track changes to every file and subdirectory. The way to create a new repository with Git is with the **init** command (short for "initialize"), which creates a special hidden directory called **.git** where Git stores the information it needs to track our project's changes. (It's the presence of a properly configured **.git** directory that distinguishes a Git repository from a regular directory.)

All Git commands consist of the command-line program **git** followed by the name of the command, so the full command to initialize a repository is **git init**, as shown in Listing 8.4.

---

5. *Learn Enough HTML to Be Dangerous* (https://www.learnenough.com/html) and *Learn Enough CSS & Layout to Be Dangerous* (https://www.learnenough.com/css-and-layout) build on this foundation to make more complicated sites.

Listing 8.4: Initializing a Git repository.

```
[website]$ git init
Initialized empty Git repository in /Users/mhartl/repos/website/.git/
[website (main)]$
```

The prompt shown in Listing 8.4 reflects both the Bash customization from Part II (Listing 6.6) and the advanced setup in Section 11.6.2, so your prompt may differ.[6] In particular, the prompt in the highlighted line in Listing 8.4 shows the name of the current Git *branch*, called **main**. Don't worry about what this means now; we'll discuss branches starting in Section 10.3.

8.2.1 Exercises

1. By running **ls -a** (Section 2.2.1), list **all** the files and directories in your **website** directory. What is the name of the hidden directory used by the Git repository? (There is one such hidden directory per project.)

2. Using the result of the previous exercise, run **ls** on the hidden directory and guess the name of the main Git configuration file. Use **cat** to dump its contents to the screen.

8.3 Our First Commit

Git won't let us complete the initialization of the repository while it's empty, so we need to make a change to the current directory. We'll make a more substantive change in a moment, but for now we'll follow a common convention and simply use **touch** to create an empty file (Listing 2.2). In this case, we're making a simple website, and the near-universal convention is to call the main page **index.html**:

```
[website (main)]$ touch index.html
```

---

6. To learn how to set up this same custom prompt using Z shell, see the Learn Enough blog post "Using Z Shell on Macs with the Learn Enough Tutorials" (https://news.learnenough.com/macos-bash-zshell).

Having created this first file, we can use the **git status** command to see the result:

```
[website (main)]$ git status
On branch main

No commits yet

Untracked files:
  (use "git add <file>..." to include in what will be committed)
        index.html

nothing added to commit but untracked files present (use "git add" to track)
```

We see here that the **index.html** file is "untracked", which means Git doesn't yet know about it. We can *add* it using the **git add** command:

```
[website (main)]$ git add -A
```

Here the **-A** option tells Git to add *all* untracked files, even though in this case there's only one. In my experience, 99% of the time you add files you'll want to add them all, so this is a good habit to cultivate, and learning how to add individual files is left as an exercise (Section 8.3.1). (By the way, the nearly equivalent command **git add .**, where the dot refers to the current directory (Section 4.3), is also common.)[7]

We can see the result of **git add -A** by running **git status** again:

```
[website (main)]$ git status
On branch main

No commits yet

Changes to be committed:
  (use "git rm --cached <file>..." to unstage)
        new file:   index.html
```

As implied by the word "unstage", the status of the file has been promoted from *untracked* to *staged*, which means the file is ready to be added to the repository. *Untracked/unstaged* and *staged* are two of the four states commonly used by Git, as

---

7. In the rare cases where the two differ, what you usually want is **git add -A**, and this is what's used in the official Git documentation (https://git-scm.com/docs/git-add), so that's what we go with here.

Figure 8.1: The main Git status sequence for changing a file.

shown in Figure 8.1. (Technically, untracked and unstaged are different states, but the distinction is rarely important because `git add` tracks and stages files at the same time.)

As shown in Figure 8.1, after putting changes in the staging area we can make them part of the local repository by *committing* them using `git commit`. (We'll cover the final step from Figure 8.1, `git push`, in Section 9.3.) Most uses of `git commit` use the command-line option `-m` to include a *message* indicating the purpose of the commit (Box 8.3). In this case, the purpose is to initialize the new repository, which we can indicate as follows:

```
[website (main)]$ git commit -m "Initialize repository"
[main (root-commit) 44c52d4] Initialize repository
 1 file changed, 0 insertions(+), 0 deletions(-)
 create mode 100644 index.html
```

(I've shown my output here for completeness, but your details will vary.)

Box 8.3: Committing to Git

By design, Git requires every commit to include a *commit message* describing the purpose of the commit. Typically, this takes the form of a single line, usually limited to around 72 characters, with an optional longer message if desired (Section 11.2.3). Although conventions for commit messages vary (as humorously depicted in the xkcd comic strip "Git Commit" (https://m.xkcd.com/1296/)), the style adopted in this tutorial is to write commit messages in the *present tense* using the *imperative mood*, as in "Initialize repository" rather than "Initializes repository" or "Initialized repository". The reason for this convention is that Git models commits as a series of text transformations, and in this context it makes sense to describe what each commit *does* instead of what it did. Moreover, this usage agrees with the convention followed by the commit messages generated by Git commands themselves (e.g., "merge" rather than "merges" or "merged"). For more information, see the GitHub article "Shiny new commit styles" (https://github.blog/2011-09-06-shiny-new-commit-styles/).

At this point, we can use **git log** to see a record of our commit:

```
[website (main)]$ git log
commit 44c52d432d294ef52bae5535dc6dcb0993175a04 (HEAD -> main)
Author: Michael Hartl <michael@michaelhartl.com>
Date:   Thu Apr 1 10:30:38 2021 -0700

    Initialize repository
```

The commit is identified by a *hash*, which is a unique string of letters and numbers that Git uses to label the commit and which lets Git retrieve the commit's changes. In my case, the hash appears as

```
44c52d432d294ef52bae5535dc6dcb0993175a04
```

but since each hash is unique your result will differ. The hash is often referred to as a "SHA" (pronounced *shah*) because of the acronym for the Secure Hash Algorithm used to generate it. We'll put these SHAs to use in Section 10.4, and several more advanced Git operations require them as well.

8.3.1 Exercises

1. Using the **touch** command, create empty files called **foo** and **bar** in your repository directory.
2. By using **git add foo**, add **foo** to the staging area. Confirm with **git status** that it worked.
3. Using **git commit -m** and an appropriate message, add **foo** to the repository.
4. By using **git add bar**, add **bar** to the staging area. Confirm with **git status** that it worked.
5. Now run **git commit** *without* the **-m** option. Use your Vim knowledge (Section 5.1) to add the message "Add bar", save, and quit.
6. Using **git log**, confirm that the commits made in the previous exercises worked correctly.

8.4 Viewing the Diff

It's often useful to be able to view the changes represented by a potential commit before making it. To see how this works, let's add a little bit of content to **index.html** by redirecting the output of **echo** (Section 2.1) to make a "hello, world" page:

```
[website (main)]$ echo "hello, world" > index.html
```

Recall from Section 2.1 that the Unix **diff** utility lets us compare two files **foo** and **bar** by typing

```
$ diff foo bar
```

Git has a similar function, **git diff**, which by default just shows the difference between the last commit and unstaged changes in the current project:

```
[website (main)]$ git diff
diff --git a/index.html b/index.html
index e69de29..4b5fa63 100644
--- a/index.html
+++ b/index.html
@@ -0,0 +1 @@
+hello, world
```

Because the content added in Section 8.3 was empty, here the diff appears simply as an addition:

```
+hello, world
```

We can commit this change by passing the **-a** option (for "all") to **git commit**, which arranges to commit all the changes in currently existing files (Listing 8.5).

Listing 8.5: Committing changes to all modified files.

```
[website (main)]$ git commit -a -m "Add content to index.html"
[main 64f6529] Add content to index.html
 1 file changed, 1 insertion(+)
```

Note that the **-a** option includes changes only to files already added to the repository, so when there are new files it's important to run **git add -A** as in Section 8.3 to make sure they're added properly. It's easy to get in the habit of running **git commit -a** and forget to add new files explicitly; learning how to deal with this situation is left as an exercise (Section 8.4.1).

Having added and committed the changes, there's now no diff:

```
[website (main)]$ git diff
[website (main)]$
```

(In fact, simply adding the changes is sufficient; running **git add -A** would also lead to there being no diff. To see the difference between staged changes and the previous version of the repo, use **git diff --staged**.)

We can confirm that the change went through by running **git log**:

```
[website (main)]$ git log
commit 64f6529494cb0e193f05b0da75702feef854e176
Author: Michael Hartl <michael@michaelhartl.com>
Date:   Thu Apr 1 10:33:24 2021 -0700

    Add content to index.html

commit 44c52d432d294ef52bae5535dc6dcb0993175a04
Author: Michael Hartl <michael@michaelhartl.com>
Date:   Thu Apr 1 10:30:38 2021 -0700

    Initialize repository
```

8.4.1 Exercises

1. Use **touch** to create an empty file called **baz**. What happens if you run **git commit -am "Add baz"**?

2. Add **baz** to the staging area using **git add -A**, then commit with the message **"Add bazz"**.

3. Realizing there's a typo in your commit message, change **bazz** to **baz** using **git commit --amend**.

4. Run **git log** to get the SHA of the last commit, then view the diff using **git show <SHA>** to verify that the message was amended properly.

8.5 Adding an HTML Tag

We've now seen all of the major elements involved in the simplest Git workflow, so in this section and the next we'll review what we've done and see how everything fits together. We'll err on the side of making more frequent commits, representing relatively modest changes, but this isn't necessarily how you should work in real life (Box 8.4). Still, it's an excellent foundation, and it will give you a solid base on which to build your own workflow and development practices.

Box 8.4: Commitment Issues

One common issue when learning Git involves figuring out when to make a commit. Unfortunately, there's no simple answer, and real-life usage varies considerably (as illustrated in the xkcd comic strip "Git Commit" (https://m.xkcd.com/1296/)). My best advice is to make a commit whenever you've reached a natural stopping point, or when you've made enough changes that you're starting to worry about losing them. In practice, this can lead to inconsistent results, and it's common to work for a while and make a large commit and then make a minor unrelated change with a small commit. This mismatch between commit sizes can seem a little weird, but it's a difficult situation to avoid.

Many teams (including most open-source projects) have their own conventions for commits, including the practice of *squashing* commits to combine them all into one commit for convenience. (Per Box 8.2, this is exactly the kind of thing you can learn about by Googling for it.) In these circumstances, I recommend following the conventions adopted by the project in question.

More than anything, don't worry about it too much. "Git Commit" is only a slight exaggeration, and in any case deciding when to commit is the kind of thing that you'll invariably get better at with time and experience.

As in previous sections, we'll be working on the main **index.html** file. Let's start by opening this file in both a text editor and a web browser. My preferred method for doing this is at the command line using the **atom** and **open** commands (though the latter works only on macOS):

```
[website (main)]$ atom index.html
[website (main)]$ open index.html      # only on macOS; otherwise, use a GUI
```

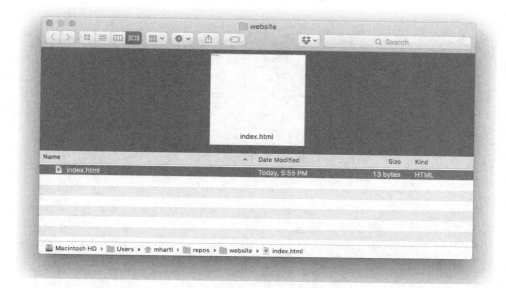

Figure 8.2: Viewing `index.html` in a filesystem browser.

If you're not on a Mac (or even if you are), you can open the directory using a graphical file browser and double-clicking the file to open it in the default browser (Figure 8.2). However you open the file, the results should appear approximately as shown in Figure 8.3 and Figure 8.4.

At this point, we're ready to make a change, which is to promote "hello, world" from ordinary text to a top-level (Level 1) heading. In HTML, the language of the World Wide Web, the way to do this is with a *tag*—in this case, the Level 1 header tag **h1**. Most browsers set **h1** tags in a large font, so the text **hello, world** should look bigger when we're done. To make the change, replace the current contents of `index.html` with the contents shown in Listing 8.6. (In this and all other examples of editing text, you'll learn more if you type in everything by hand instead of copying and pasting.)

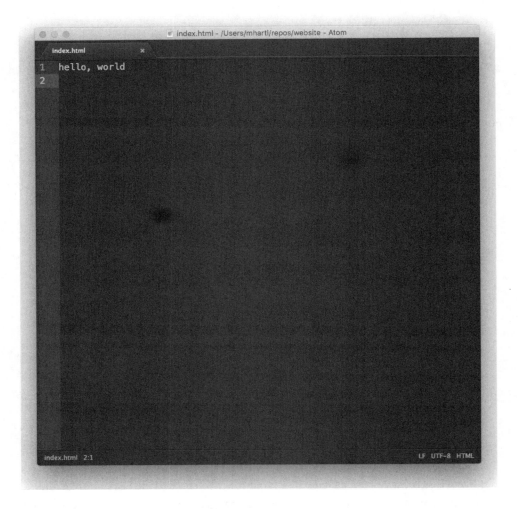

Figure 8.3: The initial HTML file opened in Atom.

Figure 8.4: The initial HTML file viewed in a web browser.

Listing 8.6: A top-level heading.

```
<h1>hello, world</h1>
```

Listing 8.6 shows the basic structure used by most HTML tags. First, there's an *opening tag* that looks like **<h1>**, where the angle brackets **<** and **>** surround the tag name (in this case, **h1**). After the content, there's a *closing tag* that's the same as the opening tag, except with an extra slash after the opening angle bracket: **</h1>**. (Note that, as with addresses on the World Wide Web, this is a *slash*, not a *backslash*— a common confusion humorously referenced in the xkcd comic strip "Trade expert" (https://m.xkcd.com/727/).)

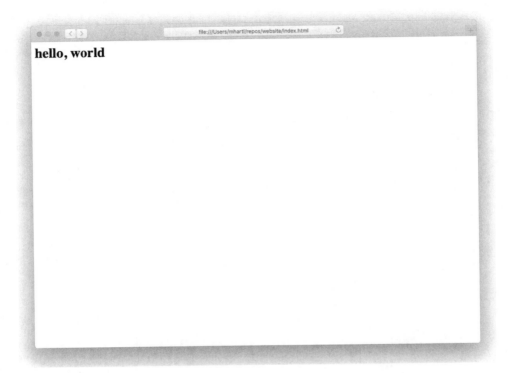

Figure 8.5: The result of adding an **h1** tag.

Upon refreshing the web browser, the index page should appear something like Figure 8.5. As promised, the font size of the text for the top-level heading is bigger (and bolder, too).

As before, we'll run **git status** and **git diff** to learn more about what we're going to commit to Git, though with experience you'll come to run these commands only when necessary. The status simply indicates that **index.html** has been modified:

```
[website (main)]$ git status
Changes not staged for commit:
  (use "git add <file>..." to update what will be committed)
  (use "git restore <file>..." to discard changes in working directory)
        modified:   index.html

no changes added to commit (use "git add" and/or "git commit -a")
```

Meanwhile, the diff shows that one line has been deleted (indicated with **-**) and another added (indicated with **+**):

```
[website (main)]$ git diff
diff --git a/index.html b/index.html
index 4b5fa63..45d754a 100644
--- a/index.html
+++ b/index.html
@@ -1 +1 @@
-hello, world
+<h1>hello, world</h1>
```

As with the Unix **diff** utility, modified sections of code or markup are shown as close to each other as possible so that it's clear at a glance what changed.[8]

At this point, we're ready to commit our changes. In Listing 8.5 we used both the **-a** and **-m** options to commit **a**ll pending changes while adding a commit **m**essage, but in fact the two can be combined as **-am** (Listing 8.7).

Listing 8.7: Committing with **-am**.

```
[website (main)]$ git commit -am "Add an h1 tag"
```

Using the **-am** combination as in Listing 8.7 is common in idiomatic Git usage.

8.5.1 Exercises

1. The **git log** command shows only the commit messages, which makes for a compact display but isn't particularly detailed. Verify by running **git log -p** that the **-p** option shows the full diffs represented by each commit.

2. Under the **h1** tag in Listing 8.6, use the **p** tag to add a *paragraph* consisting of the line "Call me Ishmael." The result should appear as in Figure 8.6. (Don't worry if you get stuck; we'll incorporate the answer to this exercise in Section 8.6 (Listing 8.8).)

---

8. When viewing small diffs, particularly in prose, the **--color-words** option is especially useful, so if the regular diff is hard to read I recommend trying **git diff --color-words** to see the effect. (This option also works with the regular Unix **diff** program.)

8.6 Adding HTML Structure

Although the web browser correctly rendered the **h1** tag in Figure 8.5, properly formatted HTML pages have more structure than just bare **h1** or **p** tags. In particular, each page should have an **html** tag consisting of a *head* and a *body* (identified with **head** and **body** tags, respectively), as well as a "doctype" identifying the document type, which in this case is a particular version of HTML called HTML5. (Don't worry about these details now; we'll cover them in more depth in *Learn Enough HTML to Be Dangerous*.)

Applying these general considerations to **index.html** leads to the full HTML structure shown in Listing 8.8. This includes the **h1** tag from Listing 8.6 and the paragraph tag from Figure 8.6. (The **title** tag, included inside the **head** tag, is empty, but in general every page should have a title, and adding one for **index.html** is left as an exercise (Section 8.6.1).)

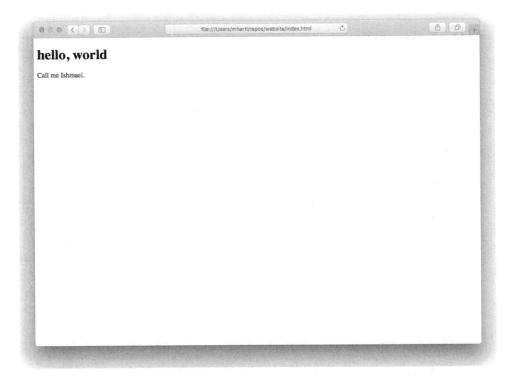

Figure 8.6: The result of adding a short paragraph.

Listing 8.8: The HTML page with added structure.

```
1   <!DOCTYPE html>
2   <html>
3     <head>
4       <title></title>
5     </head>
6     <body>
7       <h1>hello, world</h1>
8       <p>Call me Ishmael.</p>
9     </body>
10  </html>
```

Because this is a lot more content than our previous iteration (Listing 8.6), it's a good idea to go through it line by line:

1. The document type declaration

2. Opening **html** tag

3. Opening **head** tag

4. Opening and closing **title** tags

5. Closing **head** tag

6. Opening **body** tag

7. Top-level heading

8. Paragraph from the exercises (Section 8.5.1)

9. Closing **body** tag

10. Closing **html** tag

As usual, we can see the changes represented by our addition using **git diff** (Listing 8.9).

Listing 8.9: The diff for adding HTML structure.

```
[website (main)]$ git diff
diff --git a/index.html b/index.html
index 4b5fa63..afcd202 100644
--- a/index.html
+++ b/index.html
@@ -1 +1,10 @@
-<h1>hello, world</h1>
```

```
+<!DOCTYPE html>
+<html>
+  <head>
+    <title></title>
+  </head>
+  <body>
+    <h1>hello, world</h1>
+    <p>Call me Ishmael.</p>
+  </body>
+</html>
```

Despite the extensive diffs in Listing 8.9, there are hardly any user-visible differences (Figure 8.7); the only change from Figure 8.6 is a small amount of space above the top-level heading. The structure is much better, though, and brings our page nearly into compliance with the HTML5 standard. (It's not quite valid, because a

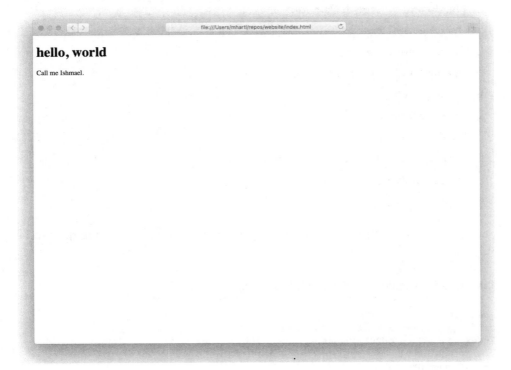

Figure 8.7: Adding HTML structure makes hardly any difference in the appearance.

nonblank page title is required by the standard; fixing this issue is left as an exercise (Section 8.6.1).)

Since we haven't added any files, using **git commit -am** suffices to commit all the changes (Listing 8.10).

Listing 8.10: The commit to add the HTML structure.

```
[website (main)]$ git commit -am "Add some HTML structure"
```

8.6.1 Exercises

1. Add the title "A whale of a greeting" to **index.html**. Browsers differ in how they display titles; the result in Safari is shown in Figure 8.8. (As of this writing, Safari

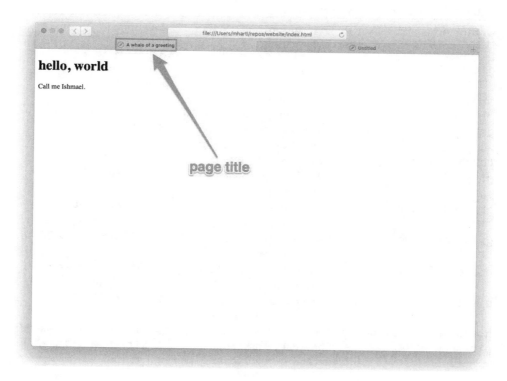

Figure 8.8: The page title displayed in a browser.

doesn't display the title unless there are at least two tabs, which is why there's a second tab in Figure 8.8.)

2. Commit the new title with a commit message of your choice. Verify using **git log -p** that the change was committed as expected.

3. By pasting the contents of Listing 8.8 into an HTML validator (https://validator .w3.org/#validate_by_input), verify that it is *not* (quite) a valid web page.

4. Using the validator, verify that the current **index.html** (with nonblank page title) *is* valid.

8.7 Summary

Important commands from this chapter are summarized in Table 8.1.

Table 8.1: Important commands from Chapter 8.

Command	Description	Example
git help	Get help on a command	$ git help push
git config	Configure Git	$ git config --global ...
mkdir -p	Make intermediate directories as necessary	$ mkdir -p repos/website
git status	Show the status of the repository	$ git status
touch <name>	Create empty file	$ touch foo
git add -A	Add all files or directories to staging area	$ git add -A
git add <name>	Add given file or directory to staging area	$ git add foo
git commit -m	Commit staged changes with a message	$ git commit -m "Add thing"
git commit -am	Stage and commit changes with a message	$ git commit -am "Add thing"
git diff	Show diffs between commits, branches, etc.	$ git diff
git commit --amend	Amend the last commit	$ git commit --amend
git show <SHA>	Show diff vs. the SHA	$ git show fb738e...

CHAPTER 9
Backing Up and Sharing

With the changes made in Chapter 8, we're now ready to push a copy of our project to a *remote repository*. This will serve as a backup of our project and its history, and will also make it easier for collaborators to work with us on our site.

We'll start by pushing our project up to *GitHub*, a site designed to facilitate collaboration with Git repositories. For repositories that are publicly available, GitHub has always been free, so we'll plan to make our website's repo public to take advantage of this. (When this tutorial was first written, GitHub charged for private repositories; in Section 11.4.1, we'll discuss an alternative that has always allowed unlimited free private repos.) Over time, releasing projects publicly on GitHub serves to build up a portfolio, which is one good reason to make as much work public as possible. There's also a Secret Reason™ for adding our repo to GitHub, which we'll get to in Section 11.4.

For reference, important commands from this chapter are summarized in Section 9.4.

9.1 Signing Up for GitHub

If you don't already have a GitHub account, you can get started by visiting the GitHub signup page (https://github.com/join) (Figure 9.1) and following the instructions. Use your technical sophistication (Box 8.2) if you get stuck.

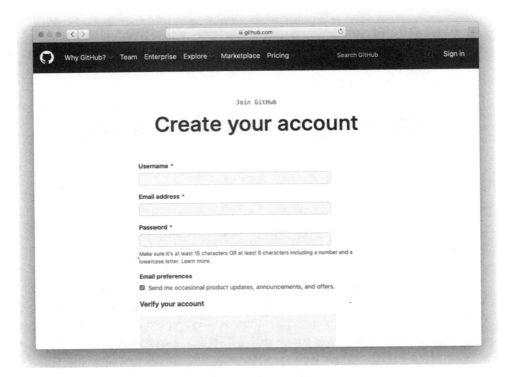

Figure 9.1: Joining GitHub.

9.2 Remote Repo

After signing up for a GitHub account, the next step is to create a remote repository. Start by selecting the menu item for adding a new repository, as shown in Figure 9.2, and then fill in the repository name ("website") and description ("A sample website for Learn Enough Git to Be Dangerous"), as shown in Figure 9.3. GitHub actively develops its user interface, so Figure 9.2, Figure 9.3, and other GitHub screenshots may not match your results exactly, but this is no cause for concern. As usual, apply your technical sophistication (Box 8.2) to resolve any discrepancies.

After clicking the button to create the repository (Figure 9.3), you should see a page like Figure 9.4 containing instructions for how to *push* your local repository up to GitHub. To get the right commands, be sure to select the setup option for HTTPS rather than SSH.

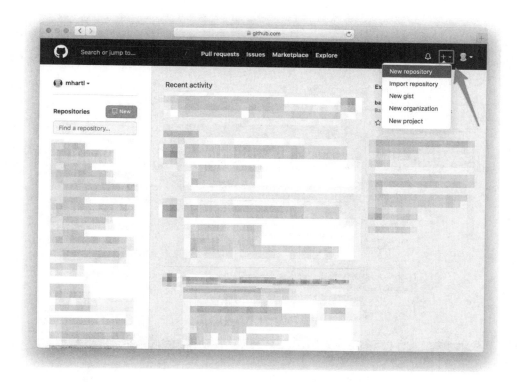

Figure 9.2: Adding a new repository at GitHub.

The exact commands in Figure 9.4 will be tailored to your personal account name and default branch name, with a template that looks like Listing 9.1. The first command sets up GitHub as the *remote origin*, the second ensures that the default branch name is **main** (which in our case does nothing since that's already its name), and the third arranges to *push* the full repository to GitHub. (The **-u** option to **git push** sets GitHub as the *upstream repository*, which means we'll be able to download any changes automatically when we run **git pull** starting in Section 11.1.) Don't worry about these details, though; you will almost always copy such commands from GitHub and probably won't ever have to figure them out on your own.

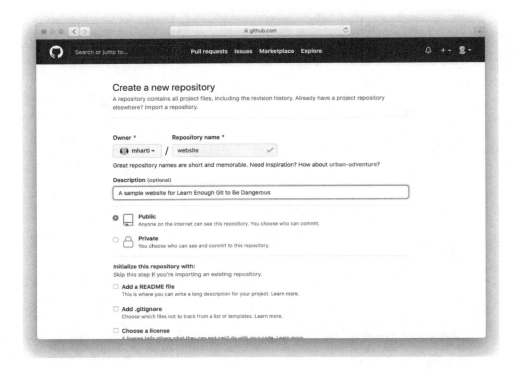

Figure 9.3: Creating a new repository.

Listing 9.1: A template for the first push to GitHub.

```
[website (main)]$ git remote add origin https://github.com/<name>/website.git
[website (main)]$ git push -u origin main
```

In Listing 9.1, you should replace **<name>** with your actual username. For example, the commands for my username, which is **mhartl**, look like this:

```
[website (main)]$ git remote add origin https://github.com/mhartl/website.git
[website (main)]$ git push -u origin main
```

After running the third command in Listing 9.1, you will be prompted to enter your username and password. The username is simply your GitHub username, but the password is *not* your GitHub password; instead, the "password" must be a *personal*

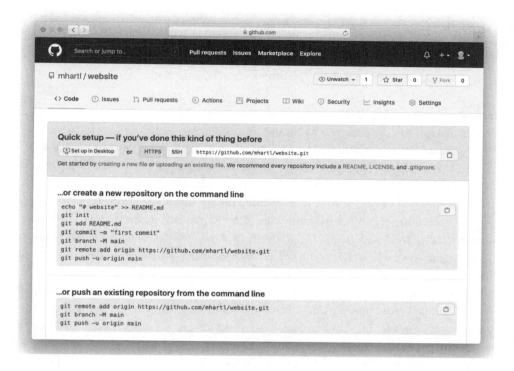

Figure 9.4: Instructions for pushing up the repo.

access token,[1] which you can create by following the instructions in the GitHub article "Creating a personal access token" (https://docs.github.com/en/authentication/ keeping-your-account-and-data-secure/creating-a-personal-access-token).[2] I suggest selecting "No expiration" for the token expiration, and also be sure to select "repo" as the scope of the token so that it works at the command line. Once you've created

---

1. The "password" used to be your actual password, but GitHub changed its interface in 2021, requiring an update to this tutorial. This is the kind of little change that happens *all the time* in tech, and it's impossible for tutorials like this one to be up-to-date at all times; as is so often the case, technical sophistication (Box 8.2) is the only general answer for dealing with issues of this kind.

2. As of this writing, the article is at https://docs.github.com/en/authentication/keeping-your-account- and-data-secure/creating-a-personal-access-token, but Googling "github creating personal access token" is also a good bet.

Figure 9.5: The browser reload page button.

and saved the personal access token, paste it in at the command line when prompted for a password to complete the **git push** in Listing 9.1.

As noted briefly above, we're using the HTTPS option for our repository URL (Figure 9.4), which sends data using the secure version of the HyperText Transfer Protocol (HTTP). This is the current GitHub default, but there's another version that uses so-called SSH keys (https://docs.github.com/en/authentication/connecting-to-github-with-ssh), which has the advantage of remembering your authentication status automatically. We'll stick with HTTPS for now, since it's simpler to use and configure. The biggest downside is that you have to input your personal access token every time you want to push any changes, which can be inconvenient. Luckily, there are ways to get your computer to remember, or *cache*, your credentials; see the article "Caching your GitHub credentials in Git" (https://docs.github.com/en/get-started/getting-started-with-git/caching-your-github-credentials-in-git) for information on how to set this up on your system.[3]

After executing the first **git push** as shown in Listing 9.1, you should reload the current page (using, e.g., ⌘R or the icon shown in Figure 9.5). The result should look something like Figure 9.6. If it does, you have officially shipped your first Git repository!

9.2.1 Exercises

1. On the GitHub page for your repo, click on "Commits" to see a list of your commits. Confirm that they match the results of running **git log** on your local system.

2. At GitHub, click on the commit for adding HTML structure (Listing 8.10). Verify that the diff for the commit agrees with the one shown in Listing 8.9.

---

3. As of this writing, the article is at https://docs.github.com/en/get-started/getting-started-with-git/caching-your-github-credentials-in-git, but Googling "caching your github credentials in git" is also a good bet.

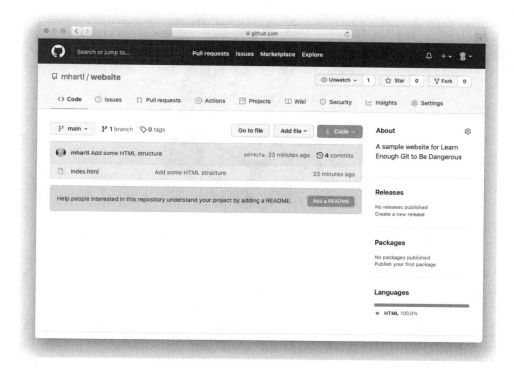

Figure 9.6: The remote repository at GitHub.

3. In honor of shipping your first Git repo, drink a celebratory beverage of your choice (Figure 9.7).[4]

9.3   Adding a README

Now that we've pushed up our repository, let's add a second file and practice the **add**, **commit**, and **push** sequence shown in Figure 8.1. You may have noticed in Figure 9.6 that GitHub encourages the presence of a README file via the note "Help people interested in this repository understand your project by adding a README."

---

4. Image courtesy of retroclipart/123rf.

Figure 9.7: Shipping a project often calls for a celebratory beverage.

Such a file literally asks the viewer to "READ ME", *à la* the DRINK ME bottle from *Alice's Adventures in Wonderland* (Figure 9.8),[5] and it's a good practice to include one.

Figure 9.6 shows a green **Add a README** button that GitHub includes to make it easy to add a README file through the web interface, but we'll follow the common (and more instructive) practice of adding it by hand locally and then pushing it up. When it comes to rendering and displaying READMEs, GitHub supports several common formats, but my favorite format for short documents like READMEs is Markdown, the lightweight markup language introduced in Section 6.2.

We can get started by opening **README.md** in Atom (or any other text editor), where the **.md** extension identifies the file as Markdown:

```
[website (main)]$ atom README.md
```

We can then fill it with the content shown in Listing 9.2.

---

5. *Alice's Adventures in Wonderland* original illustrations by John Tenniel. Colorized image courtesy of The Print Collector/Alamy Stock Photo.

Figure 9.8: Alice would know to read a README file.

Listing 9.2: The contents of the README file.
~/repos/website/README.md

```
# Sample Website

This is a sample website made as part of [*Learn Enough Git to Be
Dangerous*](https://www.learnenough.com/git-tutorial), possibly the greatest
beginner Git tutorial in the history of the Universe. You should totally [
check it out](https://www.learnenough.com/git-tutorial), and be sure to [join
the email list](https://www.learnenough.com/#email_list) and
[follow @learnenough](http://twitter.com/learnenough) on Twitter.

After finishing *Learn Enough Git to Be Dangerous*, you'll know enough Git
to be *dangerous*. This means you'll be able to use Git to track changes in
your projects, back up data, share your work with others, and collaborate
with programmers and other users of Git.
```

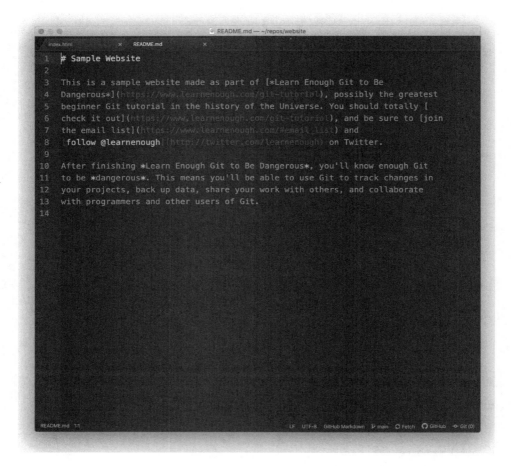

Figure 9.9: The README file viewed in Atom.

The result in Atom appears as shown in Figure 9.9. As mentioned in Section 6.2.2, Atom includes a Markdown previewer via the Packages menu item shown in Figure 9.10,[6] which (after resizing the window) results in the preview shown in Figure 9.11.[7]

---

6. If you get the error message "Previewing Markdown failed r.trim is not a function", try running the Toggle Preview keyboard shortcut from Figure 9.10 a few times to see if that helps.

7. Atom comes with a built-in Markdown previewer, but recall from Section 7.5 that editors such as Sublime Text often have installable Markdown Preview packages as well.

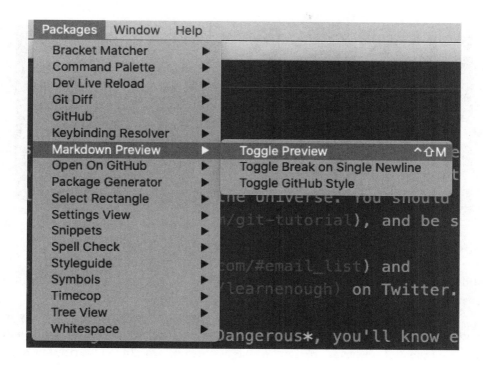

Figure 9.10: The Packages menu item for toggling the Markdown preview.

Now that we've created the **README.md** file, we're ready to add it to our Git repository and push it up. We can't just run **git commit -am** because **README.md** isn't currently in the repository, so we have to add it first:

```
[website (main)]$ git add -A
```

(As noted in Section 8.3.1, we could also run **git add README.md**, but in most cases we want to add all the new files, so I suggest getting in the habit of running **git add -A** unless there's a specific reason not to.) Then we commit as usual:

```
[website (main)]$ git commit -m "Add README file"
```

By the way, there's no harm in including **-a** via the **-am** combination shown in Listing 8.7 (and despite the redundancy I often do so out of habit), so this could just as easily read **git commit -am "Add a README file"**. (The call to **git add** is

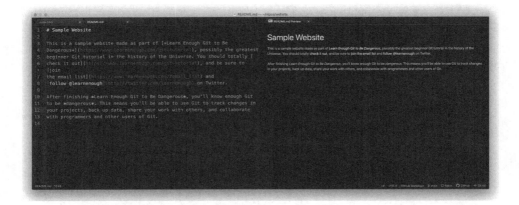

Figure 9.11: The resized Atom window with a Markdown preview.

still necessary, though; recall from Section 8.4 that **git commit -a** by itself commits changes only to files that Git is already tracking and have been modified.)

Having added the file to the repository and made a commit, we're now ready to push up to GitHub. Recall from Listing 9.1 that the first occurrence of **git push** included the "set upstream" option **-u**, the destination **origin**, and **main**, but once these are set up we can omit all those details and just **push**, like this:

```
[website (main)]$ git push
```

The result of this is to push up the new README to the remote repository, which means that we've completed the full sequence shown in Figure 8.1. In this case, GitHub uses the **.md** extension to identify the file as Markdown, converting it to HTML for easy viewing,[8] as shown in Figure 9.12.

9.3.1 Exercises

1. Using the Markdown shown in Listing 9.3, add a line at the end of the README with a link to the official Git documentation.

---

8. This involves converting the **#** in Listing 9.2 to a top-level heading (the **h1** we first saw in Section 8.5) and converting each Markdown link of the form **[content](address)** into an HTML *anchor* tag **a**, which we'll meet in Section 10.3.

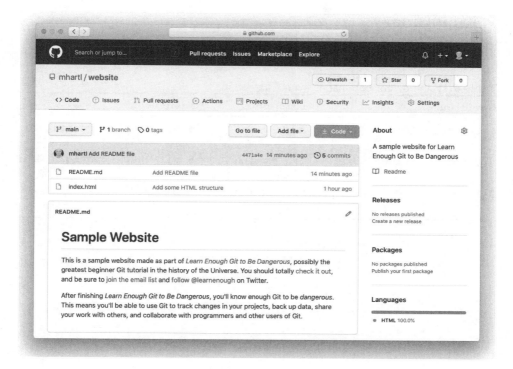

Figure 9.12: The README file at GitHub.

2. Commit your change with an appropriate message (Box 8.3). You don't have to run **git add**. Why not?

3. Push your change to GitHub. By refreshing your browser, confirm that the new line has been added to the rendered README. Click on the "official Git documentation" link to verify that it works.

Listing 9.3: Markdown code for adding a link to the official Git documentation.
~/repos/website/README.md

```
For more information on Git, see the
[official Git documentation](https://git-scm.com/).
```

9.4 Summary

Important commands from this chapter are summarized in Table 9.1.

Table 9.1: Important commands from Chapter 9.

Command	Description	Example
git remote add	Add remote repo	$ git remote add origin
git push -u <loc> 	Push branch to remote	$ git push -u origin main
git push	Push to default remote	$ git push

CHAPTER 10

Intermediate Workflow

In this chapter, we'll practice and extend the basic workflow introduced in Section 9.3. This will include adding a new directory to our project, learning how to tell Git to ignore certain files, how to *branch* and *merge*, and how to recover from errors. Rather than providing an encyclopedic coverage of Git's many commands, our focus is on covering practical techniques used every day by software developers and other users of Git.

For reference, important commands from this chapter are summarized in Section 10.5.

10.1 Commit, Push, Repeat

We'll start by adding an image to our site, which involves making a change to an existing file (**index.html**) while adding a new file in a new directory. The first step is to make a directory for images:

```
[website (main)]$ mkdir images
```

Next, download the image shown in Figure 10.1[1] to the local directory using **curl**:

```
$ curl -o images/breaching_whale.jpg \
>     -L https://cdn.learnenough.com/breaching_whale.jpg
```

---

1. Image courtesy of GUDKOV ANDREY/Shutterstock.

Figure 10.1: An image to include in our website.

(Note here that you should type the backslash character \ in the first line, but you *shouldn't* type the literal angle bracket **>** in the second line. The \ is used for a *line continuation*, and after hitting return the **>** will be added automatically by your shell program.)

We're now ready to include the image in our index page using the *image tag* **img**. This is a new kind of HTML tag; before we had opening and closing tags like

```
<p>content</p>
```

but the image tag is different. Unlike tags like **h1** and **p**, the **img** tag is a *void element* (also called a *self-closing tag*), which means that it starts with **<img** and ends with **>**:

```
<img src="path/to/file" alt="Description">
```

Note that **img** has no content between tags because there's no "between"; instead, it has a path to the *source* of the image, indicated by **src**. An alternate syntax uses **/>** instead of **>** in order to conform to constraints of XML, a markup language related to HTML:

```
<img src="path/to/file" alt="Description" />
```

You might sometimes see this syntax instead of the plain **>**, but in HTML5 the two are exactly equivalent.

By the way, in the example above the path **path/to/file** is *meta*, meaning that it talks *about* the path rather than referring to the literal path itself. In such cases, it's important to use the actual path to the file. (Successfully navigating such meta usage is a good sign of increasing technical sophistication (Box 8.2).) In this case, the path is **images/breaching_whale.jpg**, so the **img** tag in **index.html** should appear as shown in Listing 10.1. (This image tag is actually missing something important, which we'll add in Section 11.2.)

Listing 10.1: Adding an image to the index page.
~/repos/website/index.html

```
<!DOCTYPE html>
<html>
  <head>
    <title>A whale of a greeting</title>
  </head>
  <body>
    <h1>hello, world</h1>
    <p>Call me Ishmael.</p>
    <img src="images/breaching_whale.jpg">
  </body>
</html>
```

Refreshing the browser then gives the result shown in Figure 10.2. (Note that Listing 10.1 includes the **title** tag content, thereby incorporating the solution to an exercise in Section 8.6.1.)

Figure 10.2: Our website with an added image.

At this point, **git diff** confirms that the image addition is ready to go:

```
[website (main)]$ git diff index.html
diff --git a/index.html b/index.html
index 706a1be..74043f7 100644
--- a/index.html
+++ b/index.html
@@ -6,5 +6,6 @@
   <body>
     <h1>hello, world</h1>
     <p>Call me Ishmael.</p>
+    <img src="images/breaching_whale.jpg">
   </body>
 </html>
```

(If you didn't add the **title** content in Section 8.6.1, you'll see an additional line in the diff for that as well.)

On the other hand, running **git status** shows that the entire **images/** directory is untracked:

```
[website (main)]$ git status
On branch main
Your branch is up to date with 'origin/main'.

Changes not staged for commit:
  (use "git add <file>..." to update what will be committed)
  (use "git restore <file>..." to discard changes in working directory)
        modified:   index.html

Untracked files:
  (use "git add <file>..." to include in what will be committed)
        images/

no changes added to commit (use "git add" and/or "git commit -a")
```

As you might guess, **git add -A** adds all untracked *directories* in addition to adding all untracked files, so we can add the image and its directory with a single command:[2]

```
[website (main)]$ git add -A
```

We then commit and push as usual:

```
[website (main)]$ git commit -m "Add an image"
[website (main)]$ git push
```

It's a good idea to get in the habit of pushing up to the remote repository frequently, as it serves as a guaranteed backup of the project while also allowing collaborators to pull in any changes (Chapter 11).

After refreshing the GitHub repository in your browser, you should be able to confirm the presence of the new file by clicking on the **images** directory link, with the results as shown in Figure 10.3.

---

2. Technically, Git tracks only files, not directories; in fact, it won't track empty directories at all, so if you want to track an otherwise empty directory you need to put a file in it. One common convention is to use a hidden file called **.gitkeep**; to create this file in an empty directory called **foo**, you could use the command **touch foo/.gitkeep**. Then **git add -A** would add the **foo** directory as desired.

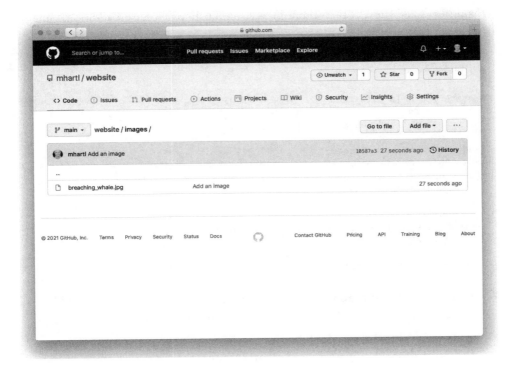

Figure 10.3: The new images directory on GitHub.

10.1.1 Exercises

1. Click on the image link at GitHub to verify that the **git push** succeeded.

2. At this point, the number of commits is large enough that the output of **git log -p** is probably too big to fit in your terminal window. Confirm that running **git log -p** drops you into a **less** interface for easier navigation.

3. Use your knowledge of **less** commands (Table 3.1) to search for the commit that added the HTML **DOCTYPE**. What is the SHA of the commit?

10.2   Ignoring Files

A frequent issue when dealing with Git repositories is coming across files you *don't* want to commit. These include files containing secret credentials, configuration files that aren't shared across computers, temporary files, log files, etc.

For example, on macOS a common side effect of using the Finder to open directories is the creation of a hidden file called `.DS_Store`.[3] This side effect is so common that more recent versions of Git actually ignore `.DS_Store` automatically, but we can simulate such a side effect by using **touch** to create a sample `.unwanted_DS_Store` file as follows:

```
[website (main)]$ touch .unwanted_DS_Store
```

This file now shows up in the status:

```
[website (main)]$ git status
On branch main
Your branch is up to date with 'origin/main'.

Untracked files:
  (use "git add <file>..." to include in what will be committed)
        .unwanted_DS_Store
nothing added to commit but untracked files present (use "git add" to track)
```

This is annoying, as we have no need to track this file, and indeed when collaborating with other users it could easily cause conflicts (Section 11.2) down the line.

In order to avoid this annoyance, Git lets us *ignore* such files using a special hidden configuration file called `.gitignore`. To ignore `.DS_Store`, create a file called `.gitignore` using your favorite text editor and then fill it with the contents shown in Listing 10.2.

Listing 10.2:   Configuring Git to ignore a file.
~/repos/website/.gitignore

```
.unwanted_DS_Store
```

---

3. This happened to me when I ran **open images/** while writing Section 10.1, which is what reminded me I should cover it here.

After saving the contents of Listing 10.2, the status now picks up the newly added **.gitignore** file, but it *doesn't* list the **.DS_Store** file, thereby confirming that it's being ignored:

```
[website (main)]$ git status
On branch main
Your branch is up to date with 'origin/main'.

Untracked files:
  (use "git add <file>..." to include in what will be committed)
      .gitignore
nothing added to commit but untracked files present (use "git add" to track)
```

This is an excellent start, but it would be inconvenient if we had to add the name of every file we want to ignore. For instance, the Vim text editor (covered briefly in Section 5.1) sometimes creates *temporary files* whose names involve appending a tilde ~ to the end of the normal filename, so you might be editing a file called **foo** and end up with a file called **foo~** in your directory. In such a case, we would want to ignore *all* files ending in a tilde. To support this case, the **.gitignore** file also lets us use *wildcards*, where the asterisk * represents "anything":[4]

```
*~
```

Adding the line above to **.gitignore** would cause all temporary Vim files to be ignored by Git. We can also add directories to **.gitignore**, so that, e.g.,

```
tmp/
```

would arrange to ignore all files in the **tmp/** directory.

Git ignore files can get quite complicated, but in practice you can build them up over time by running **git status** and looking for any files or directories you don't want to track, and then adding a corresponding pattern to the **.gitignore** file. In addition, many systems (such as the Ruby on Rails (https://rubyonrails.org/) web framework and the Softcover (https://www.softcover.io/) publishing platform)

---

4. Wildcards were discussed in Section 2.2 in the context of the **ls** command, as in **ls *.txt**.

generate a good starting .**gitignore** file for you.[5] See Chapter 1 of the *Ruby on Rails Tutorial* (https://www.railstutorial.org/book) for more information.

10.2.1   Exercises

1. Commit the .**gitignore** file to your repository. *Hint*: Running **git commit -am** isn't enough. Why not?

2. Push your commit up to GitHub and confirm using the web interface that the push succeeded.

10.3   Branching and Merging

One of the most powerful features of Git is its ability to make *branches*, which are effectively complete self-contained copies of the project source, together with the ability to *merge* one branch into another, thereby incorporating the changes into the original branch. The best thing about a branch is that you can make your changes to the project in isolation from the main copy of the code, and then merge your changes in only when they're done. This is especially helpful when collaborating with other users (Chapter 11); having a separate branch lets you make changes independently from other developers, reducing the risk of accidental conflicts.

We'll use the addition of a second HTML page, an "About page", as an example of how to use Git branches. Our first step is to use **git checkout** with the **-b** option, which makes a new branch called **about-page** and checks it out at the same time, as shown in Listing 10.3.

Listing 10.3: Checking out and creating the **about-page** branch.

```
[website (main)]$ git checkout -b about-page
[website (about-page)]$
```

The prompt in Listing 10.3 includes the new branch name for convenience, which is a result of the optional advanced setup in Section 11.6.2, so your prompt may differ.

---

5. This common practice is further evidence of the ubiquity of Git—at this point, many projects simply assume you're using it.

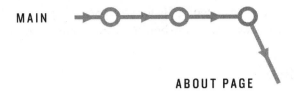

Figure 10.4: Branching off the **main** branch.

Now that we've checked out the new **about-page** branch, we can visualize our repository as shown in Figure 10.4. The main repository evolution is a series of commits, and the branch effectively represents a copy of the repo at the time the branch was made.[6] Our plan is to make a series of changes on the **about-page** branch, and then incorporate the changes back into the **main** branch using **git merge**.

We can view the current branches using the **git branch** command:

```
[website (about-page)]$ git branch
* about-page
  main
```

This lists all the branches currently defined on the local machine, with an asterisk ***** indicating the currently checked-out branch. (We'll learn how to list *remote* branches in Section 11.3.)

Having checked out the branch for the About page, we're now ready to start making some changes to our working directory. We'll start by making a new file called **about.html** to include some information about our project. Because we want the new page to have the full HTML structure (as in Listing 8.8), we'll copy over the **index.html** file and then edit it as necessary:

```
[website (about-page)]$ cp index.html about.html
```

If this duplication seems a little unclean, it is. For example, what if there were an error in the HTML structure of **index.html**? Having copied it over to **about.html**, we'd have to make the correction in both places. As we'll see in Section 11.3, in fact there

---

6. Of course, it would be potentially inefficient to copy all the files over to the new branch, since there's usually a lot of overlap with the old one. To avoid any unnecessary duplication, Git tracks diffs rather than actually making full copies of all files.

is an error, and we *will* have to make the correction twice. This sort of situation is annoying, and it's far better to use a *site template* that avoids unnecessary duplication. We'll start learning how to do that in *Learn Enough CSS & Layout to Be Dangerous* (https://www.learnenough.com/css-and-layout).

Throughout the rest of the tutorial, we'll be editing both **index.html** and **about.html**, so this is a good opportunity to use the preferred technique mentioned in Section 7.4 for opening a full project in a text editor. I suggest closing all current editor windows and re-opening the project as follows:

```
[website (about-page)]$ atom .
```

By doing this, we can use the "fuzzy opening" feature introduced in Section 7.4.1 to open the files of our choice. In particular, in Atom we can use ⌘P to open **about.html** and start making the necessary changes.

After opening **about.html**, fill it with the contents shown in Listing 10.4. As always, I recommend typing in everything by hand, which will make it easier to see the diffs relative to Listing 10.1. (The only possible exception is the trademark character ™, added to highlight character encoding issues, which you might have to copy and paste. On a Mac, you can get ™ using Option-2.)

Listing 10.4: The initial HTML for the About page.
~/repos/website/about.html

```
<!DOCTYPE html>
<html>
  <head>
    <title>About Us</title>
  </head>
  <body>
    <h1>About</h1>
    <p>
      This site is a sample project for the <strong>awesome</strong> Git
      tutorial <em>Learn Enough™ Git to Be Dangerous</em>.
    </p>
  </body>
</html>
```

Listing 10.4 introduces two new tags: **strong** (which most browsers render as **boldface** text) and **em** for emphasis (which most browsers render as *italicized* text).

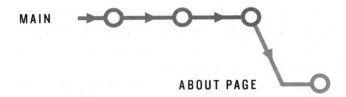

Figure 10.5: The **about-page** branch with a diff from **main**.

We're now ready to commit the initial version of the About page. Because **about.html** is a new file, we have to add it and then commit, and I sometimes like to combine these two steps using **&&** (as described in Box 4.2):

```
[website (about-page)]$ git add -A && git commit -m "Add About page"
```

At this point, the **about-page** branch has diverged from **main**, as shown in Figure 10.5.

Before merging **about-page** back in to the **main** branch, we'll make one more change. In the editor, use ⌘P or the equivalent to open **index.html** and add a *link* to the About page, as shown in Listing 10.5.

Listing 10.5: Adding a link to the About page.
~/repos/website/index.html

```
<!DOCTYPE html>
<html>
  <head>
    <title>A whale of a greeting</title>
  </head>
  <body>
    <h1>hello, world</h1>
    <a href="about.html">About this project</a>
    <p>Call me Ishmael.</p>
    <img src="images/breaching_whale.jpg">
  </body>
</html>
```

Listing 10.5 uses the important (if confusingly named) *anchor tag* **a**, which is the HTML tag for making links. This tag contains both content ("About this project") and a *hypertext reference*, or **href**, which in this case is the **about.html** file we just created.

Figure 10.6: The index page with an added link.

(Because **about.html** is on the same site as **index.html**, we can link to it directly, but when linking to external sites the href should be a fully qualified URL,[7] such as http://example.com/.)[8]

After saving the change and refreshing **index.html** in our browser, the result should appear as shown in Figure 10.6. Following the link should lead us to the About page, as seen in Figure 10.7. Note that the trademark character ™ doesn't display properly in Figure 10.7; this behavior is browser-dependent—as of this writing, the ™ symbol displays properly in Firefox and Chrome but not in Safari. We'll add code to ensure consistent behavior across all browsers in Section 11.3.

---

7. Recall that URL is short for Uniform Resource Locator, and in practice usually just means "web address".

8. Fun fact: As you can verify by visiting it, example.com is a special domain reserved for examples just like this one.

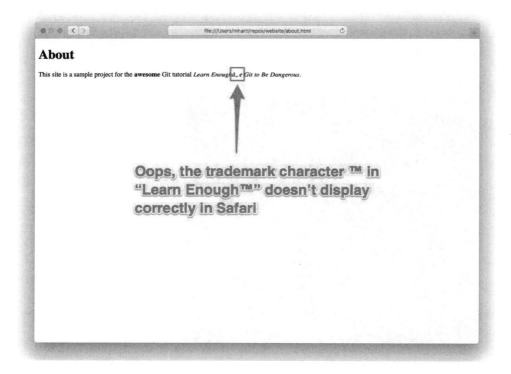

Figure 10.7: A slightly broken About page.

Having finished with the changes to **index.html**, we can make a commit as usual with **git commit -am**:

```
[website (about-page)]$ git commit -am "Add a link to the About page"
```

With this commit, the **about-page** branch now appears as in Figure 10.8.

We're done making changes for now, so we're ready to merge the About page topic branch back into the **main** branch. We can get a handle on which changes we'll be merging in by using **git diff**; we saw in Section 8.4 that this command can be used by itself to see the difference between unstaged changes and our last commit, but the same command can be used to show diffs between branches. This can take the form **git diff branch-1 branch-2**, but if you leave the branch unspecified Git

Figure 10.8: The current state of the **about-page** branch relative to **main**.

automatically diffs against the current branch. This means we can diff **about-page** vs. **main** as follows:

```
[website (about-page)]$ git diff main
```

The result in my terminal program appears as shown in Figure 10.9. On my system, the diff is too long to fit on one screen, but (as we saw with **git log** in Section 10.1.1) the output of **git diff** uses the **less** program in this case.

To incorporate the changes on **about-page** into **main**, the first step is to check out the **main** branch:

```
[website (about-page)]$ git checkout main
[website (main)]$
```

Note that, unlike the **checkout** command in Listing 10.3, here we omit the **-b** option because the **main** branch already exists.

The next step is to merge in the changes on the other branch, which we can do with **git merge**:

```
[website (main)]$ git merge about-page
Updating 5a23e6a..cad4761
Fast-forward
 about.html | 13 +++++++++++++
 index.html |  1 +
 2 files changed, 14 insertions(+)
 create mode 100644 about.html
```

At this point, our branch structure appears as in Figure 10.10.

In the present case, the **main** branch didn't change while we were working on the **about-page** branch, but Git excels even when the original branch has changed

```
●  ●  ●                        3. ~/repos/website (less)
diff --git a/about.html b/about.html
new file mode 100644
index 0000000..367dd8e
--- /dev/null
+++ b/about.html
@@ -0,0 +1,13 @@
+<!DOCTYPE html>
+<html>
+  <head>
+    <title>About Us</title>
+  </head>
+  <body>
+    <h1>About</h1>
+    <p>
+      This site is a sample project for the <strong>awesome</strong> Git
+      tutorial <em>Learn Enough™ Git to Be Dangerous</em>.
+    </p>
+  </body>
+</html>
diff --git a/index.html b/index.html
index 74043f7..b80c470 100644
--- a/index.html
+++ b/index.html
:▊
```

Figure 10.9: Diffing two branches.

Figure 10.10: The branches after merging **about-page** into **main**.

in the interim. This situation is especially common when collaborating with others (Chapter 11), but can happen even when working alone.

Suppose, for example, that we discovered a typo on **main** and wanted to fix it and push up immediately. In that case the **main** branch would change (Figure 10.11), but we could still merge in the topic branch as usual. There is a possibility that changes

Figure 10.11: The tree structure if we made a change to `main`.

on **main** would *conflict* with the merged changes, but Git is good at automatically merging content. Even when conflict is unavoidable, Git is good at marking conflicts explicitly so that we can resolve them by hand. We'll see a concrete example of this in Section 11.2.

Having merged in the changes, we can sync up the local **main** branch with the version at GitHub (called **origin/main**) as usual:

```
[website (main)]$ git push
```

Since we probably don't need the **about-page** branch any longer, we can optionally delete it, which is left as an exercise (Section 10.3.2).

10.3.1 Rebasing

The most common way to combine branches is **git merge**, but there's a second method called **git rebase** that you're likely to encounter at some point. My advice for now is: *Ignore **git rebase***. The differences between merging and rebasing are subtle, and conventions for using **rebase** differ, so I recommend using **git rebase** only when working on a team where an advanced Git user tells you to; otherwise, use **git merge** to combine the contents of two branches.

10.3.2 Exercises

1. Use the command **git branch -d about-page** to delete the topic branch. Confirm by running **git branch** that only the **main** branch is left.

2. In Listing 10.3, we used **git checkout -b** to create a branch and check it out at the same time, but it's also possible to break this into two steps. As a first step, use **git branch** to make a branch with the name **test-branch**. (This involves

passing an argument to **git branch**, as in **git branch <branch name>**.) Then confirm that the new branch exists but isn't currently checked out by running **git branch** without an argument.

3. Check out **test-branch** and use **touch** to add a file with a name of your choice, then add and commit it to the repository.

4. Check out the **main** branch and try deleting the test branch using **git branch -d** to confirm that it doesn't work. The reason is that, in contrast to the **about-page** branch, the test branch hasn't been merged into **main**, and by design **-d** doesn't work in this case. Because we don't actually want its changes, delete the test by using the related **-D** option, which deletes the branch in question even if its changes are unmerged.

10.4 Recovering from Errors

One of the most useful features of Git is its ability to let us recover from errors that would otherwise be catastrophic. The error-recovery techniques themselves can be dangerous, though, so they should always be implemented with care.

Let's consider a common scenario where we make an unintentional change to a project and want to get back to the state of the repository as of the most recent commit (a state known as **HEAD**). For example, it's a good practice to include a newline at the end of a file so that, e.g., running **tail** (Section 3.2) gives

```
[website (main)]$ tail about.html
  .
  .
  .
  </body>
</html>
[website (main)]$
```

instead of

```
[website (main)]$ tail about.html
  .
  .
  .
  </body>
</html>[website (main)]$
```

Of course, we could add such a newline using a text editor, but a common Unix idiom to accomplish the same thing is to use **echo** with the append operator **>>**:

```
[website (main)]$ echo >> about.html     # Appends a newline to about.html
```

Unfortunately, in this context it's easy to accidentally leave off one of the angle brackets and inadvertently use the *redirect* operator **>** instead (Section 2.1):

```
[website (main)]$ echo > about.html
```

Go ahead and try the command above; you will discover that the result is to overwrite **about.html** with a newline, thereby effectively wiping out its contents, as we can verify with **cat**:

```
[website (main)]$ cat about.html

[website (main)]$
```

In a regular Unix directory (Chapter 4), there would be no hope of recovering the contents of **about.html**, but in a Git repository we can undo the changes by forcing the system to check out the most recently committed version. We start by confirming that **about.html** has changed by running **git status**:

```
[website (main)]$ git status
On branch main
Your branch is up to date with 'origin/main'.

Changes not staged for commit:
  (use "git add <file>..." to update what will be committed)
  (use "git restore <file>..." to discard changes in working directory)
        modified:   about.html

no changes added to commit (use "git add" and/or "git commit -a")
```

This doesn't indicate the scope of the damage, though, which we can inspect using **git diff**:

```
$ git diff
diff --git a/about.html b/about.html
```

```
index 367dd8e..8b13789 100644
--- a/about.html
+++ b/about.html
@@ -1,13 +1 @@
-<!DOCTYPE html>
-<html>
-  <head>
-    <title>About Us</title>
-  </head>
-  <body>
-    <h1>About</h1>
-    <p>
-      This site is a sample project for the <strong>awesome</strong> Git
-      tutorial <em>Learn Enough™ Git to Be Dangerous</em>.
-    </p>
-  </body>
-</html>
+
```

Those minus signs indicate that all of the lines of content are now gone, while the plus sign at the end indicates that there's nothing left. Happily, we can undo these changes by passing the **-f** (force) option to **checkout**, which forces Git to check out **HEAD**:[9]

```
[website (main)]$ git checkout -f
```

We can then confirm that the About page has been restored:

```
[website (main)]$ git status
On branch main
Your branch is up to date with 'origin/main'.

nothing to commit, working tree clean
```

The status "working tree clean" indicates that there are no changes, and you can verify by running **cat about.html** that its contents have been restored. Phew! That was a close one. (It's worth noting that **git checkout -f** itself is potentially dangerous, as it wipes out *all* the changes you've made, so use this trick only when you're 100% sure you want to revert to **HEAD**.)

---

9. The command **git reset --hard HEAD** is equivalent, but I find the version with **checkout** to be easier to remember.

Another source of robustness against error is using branches, as described in Section 10.3. Because changes made on one branch are isolated from other branches, you can always just delete the branch if things go horribly wrong. For example, suppose we made the same **echo** mistake on a **test-branch**:

```
[website (main)]$ git checkout -b test-branch
[website (test-branch)]$ echo > about.html
```

We can fix this by committing the changes and then deleting the branch:

```
[website (test-branch)]$ git commit -am "Oops"
[website (test-branch)]$ git checkout main
[website (main)]$ git branch -D test-branch
```

Note here that we need to use **-D** instead of **-d** to delete the branch because **test-branch** is unmerged (Section 10.3.2).

A final example of recovering from error involves the common case of a bug or other defect that makes its way into a project, origins unknown. In such a case, it's convenient to be able to check out an earlier version of the repository.[10] The way to do this is to use the SHAs from the Git log (Section 8.3). For example, to restore the website project to the state right after the second commit, we would run **git log** and navigate to the beginning of the log. Because **git log** uses the **less** interface (Section 3.3), we can do this by typing **G** to go to the last line of the log. The result on my system is shown in Listing 10.6. (Because SHAs are by design unique identifiers, your values will differ.)

Listing 10.6: Viewing the SHAs in the Git log.

```
commit cad4761db5cce3544b72688329185f97a17badb3
Author: Michael Hartl <michael@michaelhartl.com>
Date:   Thu Apr 1 12:00:55 2021 -0700

    Add a link to the About page

commit 92ac96f80e9f3cbcc750d58777ca9a370aadb7f5
Author: Michael Hartl <michael@michaelhartl.com>
Date:   Thu Apr 1 11:56:21 2021 -0700
```

---

10. The most powerful way to track down such errors is **git bisect**. This advanced technique is covered in the Git documentation (https://git-scm.com/docs/git-bisect).

```
    Add About page

commit 5a23e6ac79ec1dfc5109a11780967832b43c30e3
Author: Michael Hartl <michael@michaelhartl.com>
Date:   Thu Apr 1 11:53:56 2021 -0700

    Add .gitignore

commit 10587a3a24f2eaad9659f0cc1d4bb308b169a0c2
Author: Michael Hartl <michael@michaelhartl.com>
Date:   Thu Apr 1 11:44:22 2021 -0700

    Add an image

commit 4471a4e02dfe58a229735704e4ea51ea5fc09f70
Author: Michael Hartl <michael@michaelhartl.com>
Date:   Thu Apr 1 11:19:55 2021 -0700

    Add README file

commit edf4cfa49c0b2a3bcb0f6c21f1cab4d412ce5f0d
Author: Michael Hartl <michael@michaelhartl.com>
Date:   Thu Apr 1 10:44:44 2021 -0700

    Add some HTML structure

commit eafb7bf8e1999eafa63068dabbdb05410bed512a
Author: Michael Hartl <michael@michaelhartl.com>
Date:   Thu Apr 1 10:42:48 2021 -0700

    Add an h1 tag

commit 64f6529494cb0e193f05b0da75702feef854e176
Author: Michael Hartl <michael@michaelhartl.com>
Date:   Thu Apr 1 10:33:24 2021 -0700

    Add content to index.html

commit 44c52d432d294ef52bae5535dc6dcb0993175a04
Author: Michael Hartl <michael@michaelhartl.com>
Date:   Thu Apr 1 10:30:38 2021 -0700

    Initialize repository
```

To check out the commit with the message "Add content to index.html", simply copy the SHA and check it out:

```
[website (main)]$ git checkout 64f6529494cb0e193f05b0da75702feef854e176
Note: checking out '64f6529494cb0e193f05b0da75702feef854e176'.

You are in 'detached HEAD' state. You can look around, make experimental
changes and commit them, and you can discard any commits you make in this
state without impacting any branches by performing another checkout.

If you want to create a new branch to retain commits you create, you may
do so (now or later) by using -b with the checkout command again. Example:

  git checkout -b new_branch_name

HEAD is now at 64f6529... Add content to index.html
[website ((64f6529...))]$
```

Note that the branch name in the last line has changed to reflect the value of the SHA, and Git has issued a warning that we are in a "detached HEAD" state. I recommend using this technique to inspect the state of the project and figure out any necessary changes, then check out the **main** branch to apply them:

```
[website ((64f6529...))]$ git checkout main
[website (main)]$
```

At this point, you could switch to your text editor and make any necessary changes (such as fixing a bug discovered on the earlier commit).

If all this seems a little abstract, don't worry. The main takeaways are (1) it's possible to "go back in history" to view the project at an earlier state and (2) it's tricky to make changes, so if you find yourself doing anything complicated you should ask a more experienced Git user what to do. (In particular, the exact practices in such a case could be team-dependent.)

10.4.1 Exercises

1. The **git checkout -f** trick works only with files that are staged for commit or are already part of the repository, but sometimes you want to get rid of new files as well. Using **touch**, create a file with a name of your choice, then **git add** it. Verify that running **git checkout -f** gets rid of it.

2. Like many other Unix programs, **git** accepts both "short form" and "long form" options. Repeat the previous exercise with **git checkout --force** to confirm

that the effects of **-f** and **--force** are identical. *Extra credit*: Double-check this conclusion by finding the "force" option in the output of **git help checkout**.

10.5 Summary

Important files and commands from this chapter are summarized in Table 10.1.

Table 10.1: Important commands from Chapter 10.

File/Command	Description	Example
.gitignore	Tell Git which things to ignore	$ echo .DS_store >> .gitignore
git checkout 	Check out a branch	$ git checkout main
git checkout -b 	Check out & create a branch	$ git checkout -b about-page
git branch	Display local branches	$ git branch
git merge 	Merge in a branch	$ git merge about-page
git rebase	Do something possibly weird & confusing	See "Git Commit" (https://m .xkcd.com/1296/)
git branch -d 	Delete branch (if merged)	$ git branch -d about-page
git branch -D 	Delete branch (even if unmerged) **(dangerous)**	$ git branch -D other-branch
git checkout -f	Force checkout, discarding changes **(dangerous)**	$ git add -A && git checkout -f

Chapter 11
Collaborating

Now that we've covered some of the tools needed to use Git effectively on solo projects, it's time to learn about what is perhaps Git's greatest strength: making it easier to collaborate with other people. This is especially the case when using repository hosts like GitHub (https://github.com/) or Bitbucket (https://bitbucket.org/), but it is also possible to host Git repositories on private servers (sometimes using software like GitLab (https://about.gitlab.com/) to get many GitHub-like benefits).

Because this tutorial is designed for individual readers, we won't actually be able to collaborate with others, but this chapter will explain how you can practice "collaborating" with yourself. There are many different collaboration scenarios, and they vary significantly by team and by project, so we'll focus on the important case of multiple collaborators who all have *commit rights* to a particular repo. This model is appropriate for teams where everyone can make changes without explicit approval from a project maintainer.

Open-source projects typically use a different flow involving *forking* and *pull requests*, but the details differ enough that it's best to defer to the collaboration instructions of each particular project. Consider, for example, the instructions for contributing to Ruby on Rails (https://guides.rubyonrails.org/contributing_to _ruby_on_rails.html). With the commands from this tutorial and your technical sophistication (Box 8.2), you'll be in a good position to understand and follow such instructions if you decide to get involved in contributing to open-source software or other projects under version control with Git.

For reference, important commands from this chapter are summarized in Section 11.5.

11.1 Clone, Push, Pull

As an example of a common collaboration workflow, we'll simulate the case of two developers working on the same project, in this case the simple website developed in this tutorial. We'll start with Alice (Figure 11.1)[1] working in the original **website** directory, and we'll create a second directory (**website-copy**) for her collaborator Bob (Figure 11.2).[2]

As a first step, Alice runs **git push** just to make sure all her changes are on the remote repository:

Figure 11.1: Alice, working on **website**.

---

1. *Alice's Adventures in Wonderland* original illustrations by John Tenniel. Colorized image courtesy of The Print Collector/Alamy Stock Photo.

2. Image courtesy RTRO/Alamy Stock Photo.

Figure 11.2: Bob (with son Tim), working on **website-copy**.

```
[website (main)]$ git push
```

In real life, Alice would now need to add Bob as a collaborator on the **website** repository, which she could do at GitHub by clicking on Settings > Manage Access > Invite a collaborator and then put Bob's GitHub username in the invitation box (Figure 11.3). Because we're collaborating with ourselves, we can skip this step.

Once Bob gets the notification that he's been added to the **website** repository, he can go to GitHub to get the *clone URL*, as shown in Figure 11.4. This URL lets Bob make a full copy of the repository (including its history) using **git clone**.

Ordinarily, Bob would probably use his own **repos** directory, with a project called **website** as in Alice's original, but because we're only simulating the collaboration we'll use the name **website-copy** for clarity. In addition, when doing something

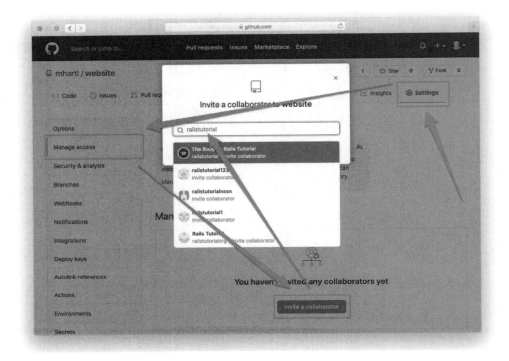

Figure 11.3: The GitHub page to add collaborators.

a little artificial like this, I like to use a temp directory called **~/tmp**,[3] so create this directory if it doesn't already exist on your system:

```
$ cd
$ mkdir tmp
```

Then **cd** to it and clone the repo to the local directory:

---

3. The idea behind a temp directory is to have a place to put temporary files that won't necessarily persist for long. Many operating systems have a system-wide temp directory (often called **/tmp**), but I also like to have one under my home directory for personal use.

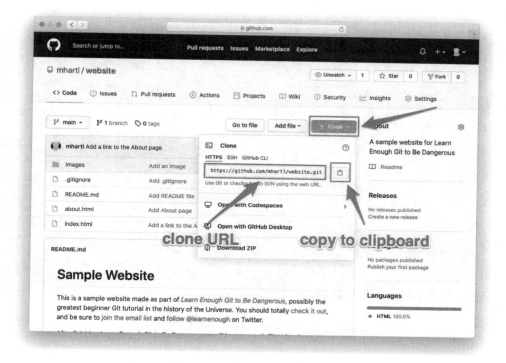

Figure 11.4: Finding the clone URL at GitHub.

```
[~]$ cd tmp/
[tmp]$ git clone <clone URL> website-copy
Cloning into 'website-copy'...
[tmp]$ cd website-copy/
```

Here we've included the argument **website-copy** to **git clone**, thereby showing how to use a different name than the original repo, but usually you just run **git clone <clone URL>**, which uses the default repo name (in this case, **website**).

Now we're ready to open the copy of the project and start making edits:

```
[website-copy (main)]$ atom .
```

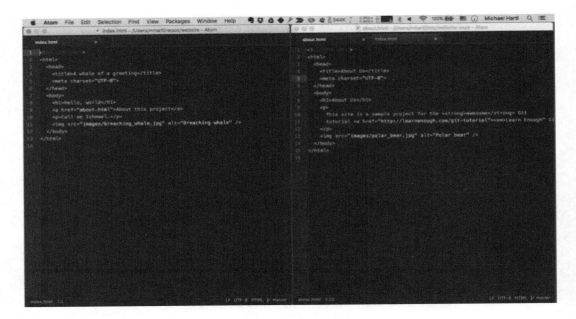

Figure 11.5: The `website` and `website-copy` editors running side by side.

For the purposes of this exercise, I recommend placing the editor windows for `website` and `website-copy` side by side, as shown in Figure 11.5.

To begin the collaboration, we'll have Bob make a change to the site by wrapping the tutorial title on the About page in a link, like this:

```
<a href="https://www.learnenough.com/git-tutorial">…</a>
```

Here the ellipsis ... represents the full title of the tutorial, *Learn Enough Git to Be Dangerous*. The resulting line is too long to display here, but we can wrap it, as shown in Figure 11.6, with the result as shown in Figure 11.7.

If we look at the diff using `git diff`, we see the wrapped line (Figure 11.8), which appears in a browser as shown in Figure 11.9.

Having added the link, Bob can commit his changes and push up to the remote repository:

Reload	^⌥⌘L
Toggle Full Screen	^⌘F
Panes	▶
Developer	▶
Increase Font Size	⇧⌘=
Decrease Font Size	⇧⌘−
Reset Font Size	⌘0
Toggle Soft Wrap	
Toggle Command Palette	⇧⌘P
Toggle Tree View	⌘\

Figure 11.6: Toggling soft wrap in Atom.

```
[website-copy (main)]$ git commit -am "Add link to tutorial title"
[website-copy (main)]$ git push
```

At this point, Bob might send Alice a notification that there's a change ready, or Alice might just be diligent about checking for changes. In either case, Alice can get the changes from the remote origin by running **git pull**. I suggest opening up a new tab in your terminal window for Alice's directory (as shown in Figure 11.10) and then pull as follows:

```
[website (main)]$ git pull
remote: Enumerating objects: 5, done.
remote: Counting objects: 100% (5/5), done.
remote: Compressing objects: 100% (1/1), done.
remote: Total 3 (delta 2), reused 3 (delta 2), pack-reused 0
Unpacking objects: 100% (3/3), 336 bytes | 168.00 KiB/s, done.
```

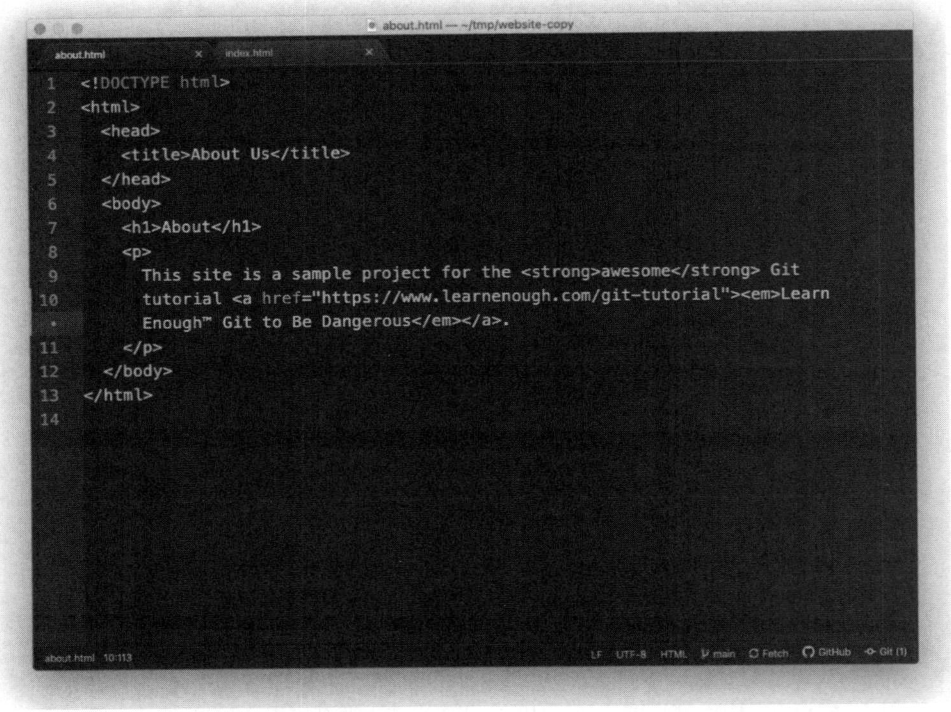

Figure 11.7: The About page with soft wrap activated.

```
From https://github.com/mhartl/website
   cad4761..9a9cecf  main          -> origin/main
Updating cad4761..9a9cecf
Fast-forward
 about.html | 2 +-
 1 file changed, 1 insertion(+), 1 deletion(-)
```

With that, Alice's project should have Bob's commit, and her copy of the About page should be identical to Figure 11.9. (Checking that Bob's commit is present in the log is left as an exercise.)

11.1.1 Exercises

1. As Alice, run **git log** to verify that the commit was pulled down correctly. Double-check the details using **git log -p**.

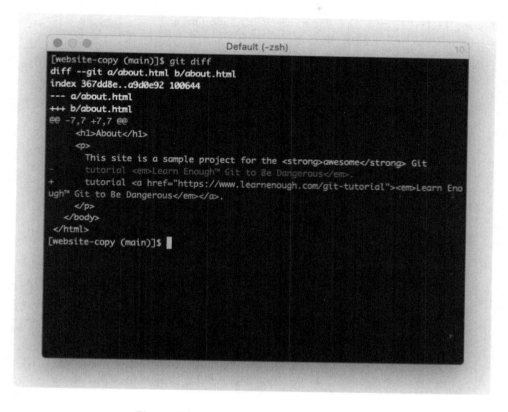

Figure 11.8: The diff with a wrapped line.

2. The whale picture added in Listing 10.1 (Figure 10.1) requires attribution under the Creative Commons Attribution-NoDerivs 2.0 Generic license. As Alice, link the image to the original attribution page, as shown in Listing 11.1. Commit the result and push to GitHub.

3. As Bob, pull in the changes from the previous exercise. Verify by refreshing the browser and by running **git log -p** that Bob's repo has been properly updated.

Listing 11.1: Linking to the whale image's attribution page.
~/repos/website/index.html

```
.
.
.
<a href="https://www.flickr.com/photos/28883788@N04/10097824543">
  <img src="images/breaching_whale.jpg">
</a>
.
.
.
```

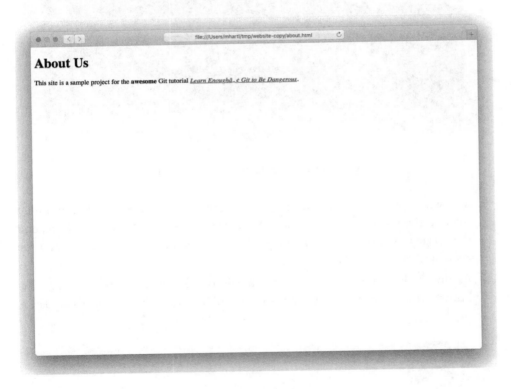

Figure 11.9: Linking the Git tutorial title on the About page.

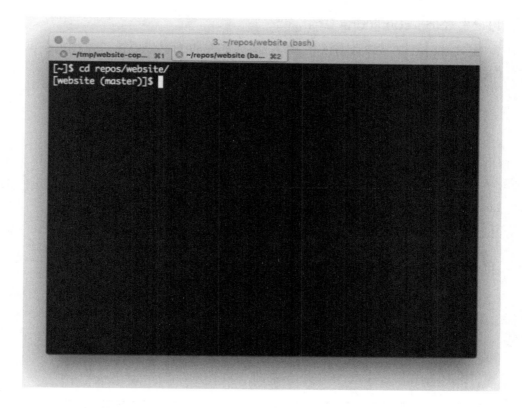

Figure 11.10: Using a new terminal tab for the original directory.

11.2 Pulling and Merge Conflicts

In Section 11.1, Alice didn't make any changes while Bob was making his commit, so there was no chance of conflict, but this is not always the case. In particular, when two collaborators edit the same file, it is possible that the changes might be irreconcilable. Git is pretty smart about merging in changes, and in general conflicts are surprisingly rare, but it's important to be able to handle them when they occur. In this section, we'll consider both non-conflicting and conflicting changes in turn.

11.2.1 Non-conflicting Changes

We'll start by having Alice and Bob make *non*-conflicting changes in the same file. Suppose Alice decides to change the top-level heading on the About page from "About" to "About Us", as shown in Listing 11.2.

Listing 11.2: Alice's change to the About page's **h1**.
~/repos/website/about.html

```
<!DOCTYPE html>
<html>
        .
        .
        .
        <h1>About Us</h1>
        .
        .

    </body>
</html>
```

After making this change, Alice commits and pushes as usual:

```
[website (main)]$ git commit -am "Change page heading"
[website (main)]$ git push
```

Meanwhile, Bob decides to add a new image (Figure 11.11)[4] to the About page. He first downloads it with **curl** as follows:

---

4. Image courtesy of Vaclav Sebek/Shutterstock.

Figure 11.11: An image for Bob to add to the About page.

```
[website-copy (main)]$ curl -o images/polar_bear.jpg \
>                           -L https://cdn.learnenough.com/polar_bear.jpg
```

(As noted in Section 10.1, you should type the backslash character \ but you *shouldn't* type the literal angle bracket **>**.) He then adds it to **about.html** using the **img** tag, as shown in Listing 11.3, with the result shown in Figure 11.12.

Listing 11.3: Adding an image to the About page.
~/tmp/website-copy/about.html

```
<!DOCTYPE html>
<html>
    .
    .
    .
    <img src="images/polar_bear.jpg" alt="Polar bear">
  </body>
</html>
```

Note that Bob has included an **alt** attribute in Listing 11.3, which is a text alternative to the image. The **alt** attribute is actually required by the HTML5 standard, and including it is a good practice because it's used by web spiders and by screen readers for the visually impaired.

Figure 11.12: The About page with an added image.

Having made his change, Bob commits as usual:

```
[website-copy (main)]$ git add -A
[website-copy (main)]$ git commit -m "Add an image"
```

When he tries to push, though, something unexpected happens, as shown in Listing 11.4.

Listing 11.4: Bob's push, rejected.

```
[website-copy (main)]$ git push
To https://github.com/mhartl/website.git
 ! [rejected]        main -> main (fetch first)
error: failed to push some refs to 'https://github.com/mhartl/website.git'
hint: Updates were rejected because the remote contains work that you do
hint: not have locally. This is usually caused by another repository pushing
hint: to the same ref. You may want to first integrate the remote changes
hint: (e.g., 'git pull ...') before pushing again.
hint: See the 'Note about fast-forwards' in 'git push --help' for details.
```

Because of the changes Alice already pushed, Git won't let Bob's push go through: As indicated by the first highlighted line in Listing 11.4, the push was rejected by GitHub. As indicated by the second highlighted line, the solution to this is for Bob to **pull**:

```
[website-copy (main)]$ git pull
```

Even though Alice made changes to **about.html**, there is no conflict because Git figures out how to combine the diffs. In particular, **git pull** brings in the changes from the remote repo and uses **merge** to combine them automatically, adding the option to add a commit message by dropping Bob into the default editor, which on most systems is Vim (Figure 11.13). (This is just one of many reasons why *Learn Enough Text Editor to Be Dangerous* (https://www.learnenough.com/text-editor) covers Minimum Viable Vim (Section 5.1).) To get the merge to go through, you can simply quit out of Vim using **:q**.

We can confirm that this worked by checking the log, which shows both the merge commit and Alice's commit from the original copy (Listing 11.5).

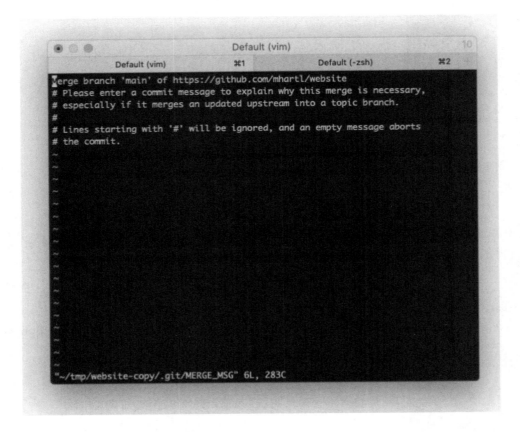

Figure 11.13: The default editor for merging from a `git pull`.

Listing 11.5: The Git log after Bob merges in Alice's changes. (Exact results will differ.)

```
[website-copy (main)]$ git log
commit 679afb8771b1893a865c3775a2786390a936db26 (HEAD -> main)
Merge: 7a69702 baafb1b
Author: Michael Hartl <michael@michaelhartl.com>
Date:   Thu Apr 1 12:28:00 2021 -0700

    Merge branch 'main' of https://github.com/mhartl/website

commit 7a6970229233346ce10cfefb3ace91b1d37c4cb2
Author: Michael Hartl <michael@michaelhartl.com>
Date:   Thu Apr 1 12:26:26 2021 -0700

    Add an image

commit baafb1bd473d553f1532267edfbbf09faf813bf2 (origin/main, origin/HEAD)
Author: Michael Hartl <michael@michaelhartl.com>
Date:   Thu Apr 1 12:25:18 2021 -0700

    Change page heading
```

If Bob now pushes, it should go through as expected:

```
$ git push
```

This puts Bob's changes on the remote repo, which means Alice can pull them in:

```
$ git pull
```

Alice can confirm that her repo now includes Bob's changes by inspecting the Git log, which should match the results you got in Listing 11.5. Meanwhile, she can refresh her browser to see Bob's cool new ursine addition (Figure 11.14).

11.2.2 Conflicting Changes

Even though Git's merge algorithms can often figure out how to combine changes from different collaborators, sometimes there's no avoiding a conflict. For example, suppose both Alice and Bob notice that the required **alt** attribute is missing from the whale image included in Listing 10.1 and decide to correct the issue by adding one.

Figure 11.14: Confirming that Alice's repo includes Bob's added image.

First, Alice adds the **alt** attribute "Breaching whale" (Listing 11.6).

Listing 11.6: Alice's image **alt**.
~/repos/website/index.html

```
<!DOCTYPE html>
<html>
    .
    .
    .
    <a href="https://www.flickr.com/photos/28883788@N04/10097824543">
      <img src="images/breaching_whale.jpg" alt="Breaching whale">
    </a>
  </body>
</html>
```

She then commits and pushes her change:[5]

```
[website (main)]$ git commit -am "Add necessary image alt"
[website (main)]$ git push
```

---

5. Listing 11.6 and Listing 11.7 include the attribution link added in the Section 11.1.1 exercises.

Listing 11.7: Bob's image **alt**.
~/tmp/website-copy/index.html

```
<!DOCTYPE html>
<html>
    .
    .
    .
    <a href="https://www.flickr.com/photos/28883788@N04/10097824543">
      <img src="images/breaching_whale.jpg" alt="Whale">
    </a>
  </body>
</html>
```

Meanwhile, Bob adds his own **alt** attribute, "Whale" (Listing 11.7), and commits his change:

```
[website-copy (main)]$ git commit -am "Add an alt attribute"
```

If Bob tries to **push**, he'll be met with the same rejection message shown in Listing 11.4, which means he should pull—but that comes at a cost:

```
[website-copy (main)]$ git pull
remote: Enumerating objects: 5, done.
remote: Counting objects: 100% (5/5), done.
remote: Compressing objects: 100% (1/1), done.
remote: Total 3 (delta 2), reused 3 (delta 2), pack-reused 0
Unpacking objects: 100% (3/3), 415 bytes | 207.00 KiB/s, done.
From https://github.com/mhartl/website
   679afb8..81c190a  main         -> origin/main
Auto-merging index.html
```

```
CONFLICT (content): Merge conflict in index.html
Automatic merge failed; fix conflicts and then commit the result.
[website-copy (main|MERGING)]$
```

As indicated in the second highlighted line, Git has detected a merge conflict from Bob's pull, and his working copy has been put into a special branch state called **main|MERGING**.

Bob can see the effect of this conflict by viewing **index.html** in his text editor, as shown in Figure 11.15. Supposing Bob prefers Alice's more descriptive **alt** text, he can resolve the conflict by deleting all but the line with **alt="Breaching whale"**, as seen in Figure 11.16. (In fact, as seen in Figure 11.15, Atom includes two "Use me" buttons to make it easy to pick one of the options. Clicking on the bottom "Use me" button gives the same result shown in Figure 11.16.)

After saving the file, Bob can commit his change, which causes the prompt to revert back to displaying the **main** branch, and at that point he's ready to **push**:

```
[website-copy (main|MERGING)]$ git commit -am "Use longer alt attribute"
[website-copy (main)]$ git push
```

Alice's and Bob's repos now have the same content, but it's still a good idea for Alice to pull in Bob's merge commit:

```
[website (main)]$ git pull
```

Because of the potential for conflict, it's a good idea to do a **git pull** before making any changes on a project with multiple collaborators (or even just being edited by the same person on different machines). Even then, on a long enough timeline some conflicts are inevitable, and with the techniques in this section you're now in a position to handle them.

Figure 11.15: A file with a merge conflict.

11.2.3 Exercises

1. Change your default Git editor from Vim to Atom. *Hint*: Google for it. (This is an absolutely *classic* application of technical sophistication (Box 8.2): With a well-chosen Google search, you can often go from "I have no idea how to do this" to "It's done" in under 30 seconds.)

2. The polar bear picture added in Listing 11.3 (Figure 11.11) requires attribution under the Creative Commons Attribution 2.0 Generic license. As Alice, link the image to the original attribution page, as shown in Listing 11.8. Then run **git commit -a** *without* including **-m** and a command-line message. This should drop you into the default Git editor. Quit the editor *without* including a message, which cancels the commit.

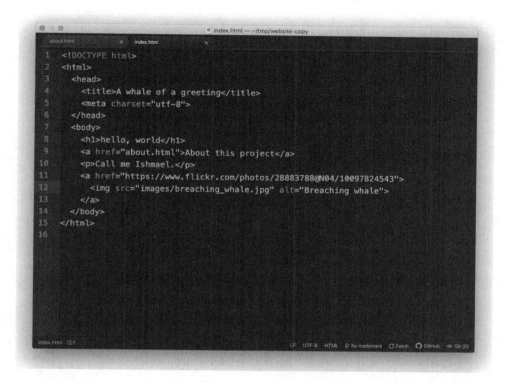

Figure 11.16: The HTML file edited to remove the merge conflict.

3. Run **git commit -a** again, but this time add the commit message "Add polar bear attribution link". Then hit return a couple of times and add a longer message of your choice. (One example appears in Figure 11.17.) Save the message and exit the editor.

4. Run **git log** to confirm that both the short and longer messages correctly appear. After pushing the changes to GitHub, navigate to the page for the commit to confirm that both the short and longer messages correctly appear.

5. As Bob, pull in the changes to the About page. Verify by refreshing the browser and by running **git log -p** that Bob's repo has been properly updated.

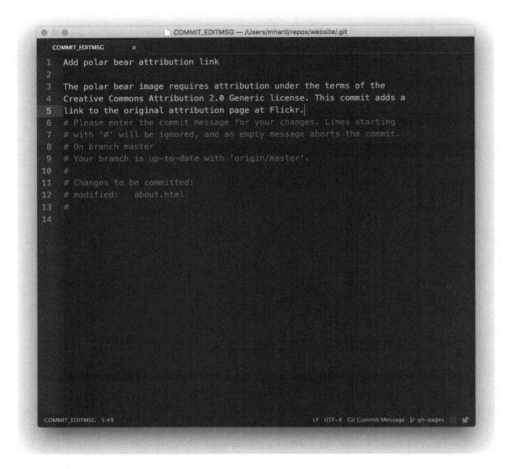

Figure 11.17: Adding a longer message in a text editor.

Listing 11.8: Linking to the polar bear image's attribution page.
~/repos/website/about.html

```
        .
        .
        .
<a href="https://www.flickr.com/photos/puliarfanita/22959238329">
  <img src="images/polar_bear.jpg" alt="Polar bear">
</a>
        .
        .
        .
```

11.3 Pushing Branches

In this section, we'll apply our newfound collaboration skills to get Alice to request a bugfix from Bob, who will make the correction and then share the result with Alice. In the process, we'll learn how to collaborate on branches other than **main**, thereby applying the material from Section 10.3 as well.

Recall from Section 10.3 that the trademark character ™ is currently broken on the About page (Figure 10.7). Alice suspects the fix for this involves adding some markup to the HTML template for the website's pages, but she's already agreed to attend a tea party (Figure 11.18),[6] so she only has time to add a couple of *HTML comments* requesting for Bob to add the relevant fix, as shown in Listing 11.9 and Listing 11.10. (We'll cover HTML comments further in *Learn Enough HTML to Be Dangerous* (https://www.learnenough.com/html).)

Figure 11.18: Alice has a tea party to attend and so asks Bob to fix the website.

---

6. *Alice's Adventures in Wonderland* original illustrations by John Tenniel. Image courtesy of The History Collection / Alamy Stock Photo.

Listing 11.9: A stub for the fix to the ™ problem.
~/repos/website/about.html

```
<!DOCTYPE html>
<html>
  <head>
    <title>About Us</title>
    <!-- Add something here to fix trademark -->
  </head>
  .
  .
  .
</html>
```

Listing 11.10: A stub to add the ™ fix to the index page.
~/repos/website/index.html

```
<!DOCTYPE html>
<html>
  <head>
    <title>A whale of a greeting</title>
    <!-- Add something here to fix trademark -->
  </head>
  .
  .
  .
</html>
```

Notice that Alice has wisely asked Bob to fix the index page as well (Listing 11.10) even though the current error only occurs on the About page. This way, any ™ or similar characters added to **index.html** will automatically work in the future. (As noted in Section 10.3, having to make such changes in multiple places is annoying, and it's also brittle and error-prone. The correct solution is to use *templates*, which we'll cover starting in *Learn Enough CSS & Layout to Be Dangerous* (https://www.learnenough.com/css-and-layout).)

Alice has decided to follow a common convention and use a separate branch for the bugfix, which in this case she calls **fix-trademark**:

```
[website (main)]$ git checkout -b fix-trademark
[website (fix-trademark)]$
```

This shows something important: It's possible to make changes to the working directory (in this case, the additions from Listing 11.9 and Listing 11.10) *before* creating a new branch, as long as those changes haven't yet been committed.

Having made the new branch for the fix, Alice can make a commit and push up the branch using **git push**:

```
[website (fix-trademark)]$ git commit -am "Add placeholders for the TM fix"
[website (fix-trademark)]$ git push -u origin fix-trademark
```

Here Alice has used exactly the same **push** syntax used in Listing 9.1 to push the repo up to GitHub in the first place, with **fix-trademark** in place of **main**.

If Alice sends Bob a note before she heads off to her tea party, Bob will know to do a **git pull** to pull in Alice's changes:

```
[website-copy (main)]$ git pull
remote: Enumerating objects: 7, done.
remote: Counting objects: 100% (7/7), done.
remote: Compressing objects: 100% (1/1), done.
```

```
remote: Total 4 (delta 3), reused 4 (delta 3), pack-reused 0
Unpacking objects: 100% (4/4), 444 bytes | 148.00 KiB/s, done.
From https://github.com/mhartl/website
 * [new branch]      fix-trademark -> origin/fix-trademark
Already up to date.
```

Bob can check his local working directory for the **fix-trademark** branch that Alice created and pushed, but it isn't there:

```
[website-copy (main)]$ git branch
* main
```

The reason is that the branch is associated with the remote **origin**, and such branches aren't displayed by default. To see it, Bob can use the **-a** option (for "all"):[7]

```
[website-copy (main)]$ git branch -a
* main
  remotes/origin/HEAD -> origin/main
  remotes/origin/fix-trademark
  remotes/origin/main
```

To start work on **fix-trademark** on his local copy, Bob just needs to check it out. By using the same name (i.e., **fix-trademark**), he arranges for it to be associated with the upstream branch on GitHub, which means that **git push** will automatically push up his changes:

---

7. In fact, **git branch --all** works, but when using Git at the command line it's more common to use the short forms of the options.

```
[website-copy (main)]$ git checkout fix-trademark
Branch fix-trademark set up to track remote branch fix-trademark from origin.
Switched to a new branch 'fix-trademark'
[website-copy (fix-trademark)]$
```

At this point, Bob can **diff** against **main** to see what he's dealing with:

```
[website-copy (fix-trademark)]$ git diff main
diff --git a/about.html b/about.html
index 173e5fe..4d4b780 100644
--- a/about.html
+++ b/about.html
@@ -2,6 +2,7 @@
 <html>
   <head>
     <title>About Us</title>
+    <!-- Add something here to fix trademark -->
   </head>
   <body>
     <h1>About Us</h1>
diff --git a/index.html b/index.html
index 024ada5..d8e946f 100644
--- a/index.html
+++ b/index.html
@@ -2,6 +2,7 @@
 <html>
   <head>
     <title>A whale of a greeting</title>
+    <!-- Add something here to fix trademark -->
   </head>
   <body>
     <h1>hello, world</h1>
```

Now all Bob has to do is actually implement the fix. If you'd like a challenging exercise in technical sophistication, try Googling around to see if you can figure out what the problem might be, and also how you might fix it. In case you'd like to do this, I'll wait here while you look…

All right, the problem is that the page doesn't have the right *character encoding* to display non–ASCII characters like ™, ®, or £. The fix involves using a tag called **meta** to tell browsers to use a character set (or **charset** for short) called UTF–8, which will let our page display anything that's part of the enormous set of Unicode characters. The result, which you would not necessarily be able to guess, appears in Listing 11.11 and Listing 11.12.

Listing 11.11: A fix for the ™ problem.
~/tmp/website-copy/about.html

```
<!DOCTYPE html>
<html>
  <head>
    <title>About Us</title>
    <meta charset="utf-8">
  </head>
    .
    .
    .
</html>
```

Listing 11.12: Adding the ™ fix to the index page.
~/tmp/website-copy/index.html

```
<!DOCTYPE html>
<html>
  <head>
    <title>A whale of a greeting</title>
    <meta charset="utf-8">
  </head>
    .
    .
    .
</html>
```

Like the **img** tag introduced in Section 10.1, **meta** is a void element and so has no closing tag.

Figure 11.19: Confirming a working trademark character.

Having made the change, Bob can confirm the fix by reloading the page in his browser, as shown in Figure 11.19.

Confident that his solution is correct, Bob can now make a commit and push the fix up to the remote server:

```
[website-copy (fix-trademark)]$ git commit -am "Fix trademark character display"
[website-copy (fix-trademark)]$ git push
```

Figure 11.20: Bob's reward for a job well-done.

With that, Bob sends a note to Alice that the fix is pushed, and heads out for some well-deserved rest (Figure 11.20).[8]

Alice, now back from her tea party, gets Bob's note and pulls in his fix:

```
[website (fix-trademark)]$ git pull
```

She refreshes her browser to confirm that the ™ character displays properly on her end of things (Figure 11.21), and then merges the changes into **main**:

---

8. Image courtesy of Maxim Safronov/Shutterstock.

Figure 11.21: Reconfirming the trademark fix before merging.

```
[website (fix-trademark)]$ git checkout main
[website (main)]$ git merge fix-trademark
[website (main)]$ git push
```

With the final **git push**, Alice arranges for the remote **main** branch on GitHub to get the fix. (Syncing up Bob's **main** branch is left as an exercise (Section 11.3.1).)

Of course, **git push** publishes the change only to a remote Git repository. Wouldn't it be nice if there were a way to confirm that the ™ character—and the rest of the website—displays correctly on the live Web?

11.3.1 Exercises

1. Bob's **main** branch doesn't currently have Alice's merge, so check out **main** as Bob and do a **git pull**. Confirm using **git log** that Alice's merge commit is now present.

2. Delete the **fix-trademark** branch locally. Do you need to use the **-D** option (Section 10.3.2), or is **-d** sufficient?

3. Delete the remote **fix-trademark** branch on GitHub. *Hint*: If you get stuck, Google for it.

11.4 A Surprise Bonus

As hinted at the end of the last section, it would be nice to be able to confirm that the new character encoding works on a live web page. But this requires knowing how to deploy a live site to the Web, and that's beyond the scope of a humble Git tutorial, right? Amazingly, the answer is no. The reason is that GitHub offers a free service called *GitHub Pages*, and *any* repository at GitHub containing static HTML is automatically available as a live website.

There is one minor prerequisite to using GitHub Pages, which is that you have to verify your email address with GitHub. Once you've done that, though, all you need to do is configure your repository to use GitHub Pages on the **main** branch, which you can do by going to the settings (Figure 11.22) and then selecting the **main** option (Figure 11.23) and saving the changes (Figure 11.24).

That's it! Our website is now available at the URL

```
https://<name>.github.io/website/
```

where **<name>** is your GitHub username. Since my username is **mhartl**, my copy of this tutorial's website is at mhartl.github.io/website/, as shown in Figure 11.25.

Note that the URL **https://<name>.github.io/website/** automatically displays **index.html**, which is the usual convention on the Web: The index page is understood to be the default, so there's no need to type it in. This is not the case with other pages, though, and if you follow the link to the About page you'll see that the filename appears in the address bar (Figure 11.26). As seen in Figure 11.27, the trademark character ™ also renders correctly on the live website, just as we hoped it would.

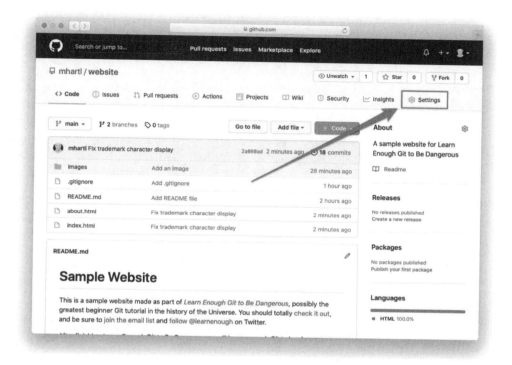

Figure 11.22: The settings for a GitHub repository.

Because static HTML pages by definition don't change from one page view to the next, GitHub can *cache* them efficiently, storing them for the next user who visits the site. This makes GitHub Pages sites both fast and cheap to serve (which is why GitHub can afford to offer them for free). You can even configure GitHub Pages to work with a custom domain, letting you replace <name>.github.io with something like www.example.com; see the free tutorial *Learn Enough Custom Domains to Be Dangerous* (https://www.learnenough.com/custom-domains) to learn how to do it. This combination of high performance and support for custom domains makes GitHub Pages suitable for production websites—for example, the Learn Enough blog (https://news .learnenough.com/) is a static website running on a custom domain at GitHub Pages.

The example website in this tutorial is really just a toy, but it's a great start, and we'll build on this foundation to make a nearly industrial-grade website in *Learn Enough HTML to Be Dangerous* and a fully industrial-grade site in *Learn Enough CSS & Layout to Be Dangerous*.

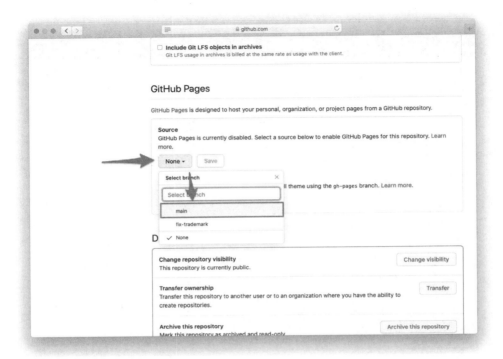

Figure 11.23: Serving our website from the **main** branch.

11.4.1 Exercises

1. On the About page, add a link back to **index.html**. Commit and push your change and verify that the link works on the production site.

2. As noted in Section 2.3, two of the most important Unix commands are **mv** and **rm**. Git provides analogues of these commands, which have the same effect on local files while also arranging to track the changes. Experiment with these commands via the following sequence: Create a file with some *lorem ipsum* text, add & commit it, rename it with **git mv** & commit, then remove it with **git rm** & commit again. Examine the results of **git log -p** to see how Git handled the operations.

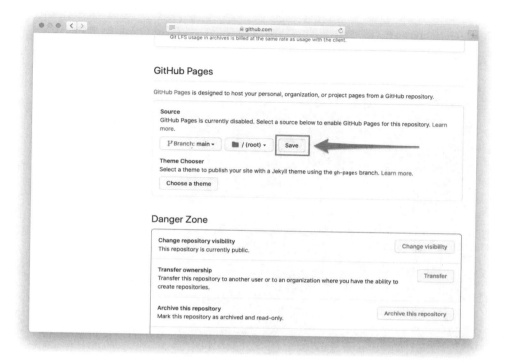

Figure 11.24: Saving the new GitHub Pages settings.

3. To practice the process of making a new Git repository, make a second project called **second_website** in the **repos** directory. Create an **index.html** file with the content "hello, again!" and follow the steps (starting in Section 8.2) needed to deploy it to the live Web.

4. Make a third, secret project called **secret_project**. Touch files called **foo**, **bar**, and **baz** in the main project directory, and then follow the steps to initialize the repository and commit the initial results. Then, to practice using a service other than GitHub, create a free private repository at Bitbucket.

11.5   Summary

Important commands from this chapter are summarized in Table 11.1.

Figure 11.25: A production website at GitHub Pages.

Figure 11.26: An explicit `about.html` in the address bar.

11.6 Advanced Setup

This section contains some optional advanced Git setup. The main features are adding an alias for checking out branches, adding the branch name to the Unix prompt, and enabling branch name tab completion. Following the steps in this section should be within your capabilities if you completed *Learn Enough Command Line to Be Dangerous* (https://www.learnenough.com/command-line) and *Learn Enough Text Editor to Be Dangerous*, but they can be tricky, so use your technical sophistication (Box 8.2) if you

Figure 11.27: The About page in production.

get stuck. If you'd rather skip these steps for now, you can proceed directly to the conclusion (Section 11.7).

Note for Mac users: The instructions below assume you are using Bash, as described in Box 2.3. To learn how to set up your Git system using Z shell instead, see the Learn Enough blog post "Using Z Shell on Macs with the Learn Enough Tutorials" (https://news.learnenough.com/macos-bash-zshell).

11.6.1 A Checkout Alias

In Chapter 8, we added global configuration settings for the name and email address (Listing 8.3) to be included automatically when making commits. Now we'll add a third config setting, an *alias* to make it easier to check out branches.

Table 11.1: Important commands from Chapter 11.

Command	Description	Example
`git clone <URL>`	Copy repo (incl. full history) to local disk	`$ git clone https:// ex.co/repo.git`
`git pull`	Pull in changes from remote repository	`$ git pull`
`git branch -a`	List all branches	`$ git branch -a`
`git checkout  `	Check out remote branch and configure for	`$ git checkout fix-trademark`

Throughout this tutorial, we've used **git checkout** to check out branches (e.g., Listing 10.3), but most experienced Git users configure their systems to use the shorter command **git co**.[9] The way to do this is with a Git *alias*: Much as the Bash aliases covered in Section 5.4 let us add commands to our Bash shell, Git aliases let us add commands to our Git system. In particular, the way to add the **co** alias is to run the command shown in Listing 11.13.

Listing 11.13: Adding an alias for **git co**.

```
$ git config --global alias.co checkout
```

In effect, this adds **co** as a new Git command, and running Listing 11.13 allows us to replace **checkout** in commands like

```
$ git checkout main
```

with the more compact **co** command, as follows:

```
$ git co main
```

For maximum compatibility with systems that don't have **co** configured, this tutorial has always used the full **checkout** command, but in real life I nearly always use **git co**.

---

9. This choice is no doubt influenced by the analogous command **svn co** used by Subversion, one of Git's main predecessors.

11.6.2 Prompt Branches and Tab Completion

In this section, we'll add two final advanced customizations. First, we'll arrange for the command-line prompt to include the name of the current branch. Second, we'll add the ability to fill in Git branch names using *tab completion* (Box 2.4), which is especially convenient when dealing with longer branch names. Both of these features come as shell scripts with the Git source code distribution, which can be downloaded as shown in Listing 11.14.

Listing 11.14: Downloading scripts for branch display and tab completion.

```
$ curl -o ~/.git-prompt.sh -L https://cdn.learnenough.com/git-prompt.sh
$ curl -o ~/.git-completion.bash \
>       -L https://cdn.learnenough.com/git-completion.bash
```

Here the **-o** flag arranges to save the files locally under slightly different names from the ones on the server, prepending a dot . so that the files are hidden (Section 2.2.1) and saving them in the home directory ~.

After downloading the scripts as in Listing 11.14, on some systems we need to make them executable, which we can do with the **chmod** command (Section 7.3):

```
$ chmod +x ~/.git-prompt.sh
$ chmod +x ~/.git-completion.bash
```

Next, we need to tell the shell about the new commands, so open up the Bash profile file in your favorite editor (which for simplicity I'll assume is Atom):

```
$ atom ~/.bashrc
```

Then add the configuration shown in Listing 11.15 to the bottom of the file. Also, make sure to delete any other lines starting with **PS1** (which you'll have to do if you modified **.bashrc** as shown in Listing 6.6).

Listing 11.15: Adding Git configuration to Bash.

`~/.bashrc`

```
.
.
.
# Git configuration
# Branch name in prompt
source ~/.git-prompt.sh
PS1='[\W$(__git_ps1 " (%s)")]\$ '
export PROMPT_COMMAND='echo -ne "\033]0;${PWD/#$HOME/~}\007"'
# Tab completion for branch names
source ~/.git-completion.bash
```

Note: The vertical dots in Listing 11.15 indicate omitted content and should not be copied literally. This is the sort of thing you can figure out using your technical sophistication (Box 8.2). Speaking of which, I have hardly any idea of what most of the code in Listing 11.15 means; part of having technical sophistication means being able to copy things from the Internet and getting them to work even when you have no idea what you're doing (Figure 11.28[10]).

Once we've saved the result of editing `.bashrc`, we have to *source* it to make the changes active (as seen in Listing 5.5):

```
$ source ~/.bashrc
```

At this point, the prompt for a Git repository's default **main** branch should look something like this:

```
[website (main)]$
```

If you skipped ahead from Section 8.1 to complete this section, you'll have to wait until Section 8.2 to see this effect. Checking that tab completion is working is left as an exercise (Section 11.6.3).

11.6.3 Exercises

1. Create a branch called **really-long-branch-name** using **git co -b**.

2. Switch back to the **main** branch using **git co**.

---

10. Image courtesy of Adam Frank/Shutterstock.

Figure 11.28: It's OK—neither does anyone else.

3. Check out the branch **really-long-branch-name** using tab completion by typing **git checkout r** ⇥ at the command-line prompt.

4. What does your prompt look like? Verify that the correct branch name appears in the prompt.

5. Check out the **main** branch using **git co m** ⇥. (This shows that tab completion works with the **co** alias set up in Listing 11.13.) What does the prompt look like now?

6. Use **git branch -d r** ⇥ to delete **really-long-branch-name**, thus verifying that tab completion works with **git branch** as well as with **git checkout**. (In fact, tab completion works with most relevant Git commands.)

11.7 Conclusion

Congratulations! You now know enough Git to be *dangerous*—which means, with
Part I and Part II, you know enough *developer tools* to be dangerous as well.

There's a lot more to learn, and if you continue down this technical path you'll
keep getting better at using Git for years to come, but with the material in this tutorial
you've got a great start. For now, you're probably best off working with what you've
got, applying your technical sophistication (Box 8.2) when necessary. Once you've
gotten a little more experience under your belt, I recommend seeking out additional
resources. Here are some suggestions for getting started:

- *Pro Git* by Scott Chacon and Ben Straub (https://git-scm.com/book/en/v2)
- Learn Git at Codecademy (https://www.codecademy.com/learn/learn-git)
- Git tutorials (https://www.atlassian.com/git/tutorials) by Atlassian (makers of Bitbucket)
- Tower Git tutorials (https://www.git-tower.com/learn/)

At this point, you are in an excellent position to collaborate with millions of
software developers around the world. You are also well on your way to becoming
a developer yourself. Regardless of your ultimate goals, you can continue improving
your dev skills with the rest of the core Learn Enough sequence:

1. ***Learn Enough Developer Tools to Be Dangerous***
   (a) Part I: *Learn Enough Command Line to Be Dangerous*
   (b) Part II: *Learn Enough Text Editor to Be Dangerous*
   (c) Part III: *Learn Enough Git to Be Dangerous* (you are here)
2. **Web Basics**
   (a) *Learn Enough HTML to Be Dangerous* (https://www.learnenough.com/html)
   (b) *Learn Enough CSS & Layout to Be Dangerous* (https://www.learnenough.com/css-and-layout)
   (c) *Learn Enough JavaScript to Be Dangerous* (https://www.learnenough.com/javascript)
3. **Application Development**
   (a) *Learn Enough Ruby to Be Dangerous* (https://www.learnenough.com/ruby)
   (b) *Ruby on Rails Tutorial* (https://www.railstutorial.org/)

(c) *Learn Enough Action Cable to Be Dangerous* (https://www.learnenough.com/action-cable) (optional)

Good luck!

APPENDIX

Development Environment

One of the most important tasks for any aspiring developer—or any technical person generally—is setting up their computer as a *development environment*, making it suitable for developing websites, web applications, and other software. This appendix to *Learn Enough Developer Tools to Be Dangerous*, also available separately as *Learn Enough Dev Environment to Be Dangerous* (https://www.learnenough.com /dev-environment-tutorial), is designed to complement the main Learn Enough sequence (https://www.learnenough.com/) and the *Ruby on Rails Tutorial* (https://www.railstutorial.org/) by putting all of the relevant material in one place.[1]

Learn Enough Dev Environment to Be Dangerous covers several options for setting up a dev environment, aimed at readers of varying levels of experience and sophistication. If you end up using the option shown in Section A.2, there is no minimum prerequisite for this tutorial other than general computer knowledge. If you want to follow the more challenging setup in Section A.3—which we recommend most readers tackle at some point—you should have a basic familiarity with the Unix command line (as covered in *Learn Enough Command Line to Be Dangerous* (https://www.learnenough.com/command-line)), and a familiarity with text editors and system configuration (as covered in *Learn Enough Text Editor to Be Dangerous* (https://www.learnenough.com/text-editor)) is recommended.

Ebook versions of this material are available for free at learnenough.com. See the online version at learnenough.com/dev-environment (https://www.learnenough .com/dev-environment) for the most up-to-date instructions on setting up a development environment.

---

1. *Learn Enough Dev Environment to Be Dangerous* was prepared with the help of Learn Enough cofounders Lee Donahoe and Nick Merwin.

A.1 Dev Environment Options

Our focus in this appendix is on installing or otherwise enabling the following four fundamental tools of software development (Figure A.1):

1. Command-line terminal ("shell")

2. Text editor

3. Version control (Git)

4. Programming languages (Ruby, etc.)

For more information on these different types of software application, see *Learn Enough Command Line to Be Dangerous*, *Learn Enough Text Editor to Be Dangerous*, *Learn Enough Git to Be Dangerous* (https://www.learnenough.com/git), and *Learn Enough Ruby to Be Dangerous* (https://www.learnenough.com/ruby).

When setting up a development environment, there are two main possibilities we recommend, listed in increasing order of difficulty:

1. Cloud IDE

2. Native OS (macOS, Linux, Windows)

Figure A.1: Typical elements of a dev environment.

If you're relatively inexperienced, we recommend starting with the cloud IDE (Section A.2), as it has the least difficult setup process.

Native System

While the IDE option is great when you're just getting started, eventually it's important to be able to develop software on your native operating system (OS). Unfortunately, setting up a fully functional native development environment can be a challenging and frustrating process[2]—likely leaving ample opportunity to exercise your technical sophistication (Box A.1)—but it is an essential rite of passage for every aspiring technical wizard.

In order to tackle this difficult challenge, in Section A.3 we'll discuss native OS setup for macOS, Linux, and Windows.

Box A.1: Technical Sophistication

Technical sophistication is the ability to independently solve technical problems. In the context of installing a development environment, this means knowing to Google the error message if something goes wrong, to try quitting and restarting an application (such as the command-line shell) to see if that fixes things, etc.

So many things can go wrong when setting up a development environment that there's rarely a general solution to the problem—you just have to keep applying your technical sophistication until you get everything to work. And if you do get stuck, don't worry too much about it—at some point or another, it happens to us all.

A.2 Cloud IDE

The easiest dev environment option is a *cloud IDE*, which is an integrated development environment in the cloud that you access using the web browser of your choice. Although easy to activate, the resulting system is an industrial-grade development

---

2. This is why it's such a bad idea to include a native dev setup at the beginning of a book like the *Ruby on Rails Tutorial* (https://www.railstutorial.org/book). It's better to get started with the core material first, and tackle the native dev environment setup later. The switch over to a cloud IDE in the third edition of the *Ruby on Rails Tutorial* was motivated by this realization.

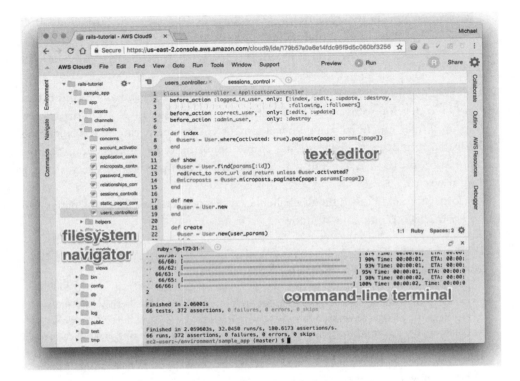

Figure A.2: The anatomy of the cloud IDE.

machine, not a toy. In addition, the cloud IDE automatically works cross-platform, since all you need is an ordinary web browser to use it (which every major OS provides).

There are several commercial options for running a cloud IDE, and as part of developing the *Ruby on Rails Tutorial* we partnered with Cloud9 (part of Amazon Web Services). The resulting environment is appropriate for Ruby on Rails web development, and as a matter of course includes all the elements mentioned in Section A.1. In particular, AWS Cloud9 comes equipped with a command-line terminal and a text editor (including a filesystem navigator), as shown in Figure A.2. Because each Cloud9 workspace provides a full working Linux system, it also automatically includes the Git version control system, as well as Ruby and several other programming languages.

Here are the steps for getting started with the cloud development environment:[3]

1. Because Cloud9 is part of Amazon Web Services (AWS), if you already have an AWS account you can just sign in.[4] To create a new Cloud9 workspace environment, go to the AWS console and type "Cloud9" in the search box.

2. If you don't already have an AWS account, you should sign up for a free account at AWS Cloud9.[5] In order to prevent abuse, AWS requires a valid credit card for signup, but the workspace is 100% free (for a year as of this writing), and your card will not be charged. You might have to wait up to 24 hours for the account to be activated, but in my case it was ready in about ten minutes.

3. Once you've successfully gotten to the Cloud9 administrative page (Figure A.3), keep clicking on "Create environment" until you find yourself on a page that looks like Figure A.4. Enter the information as shown there, then make sure to choose **Ubuntu Server** (*not* Amazon Linux) (Figure A.5). Finally, keep clicking the confirmation buttons until Cloud9 starts provisioning the IDE (Figure A.6). You may run into a warning message about being a "root" user, which you can safely ignore at this early stage. (If you're feeling up to it, you can implement the preferred method, called an Identity and Access Management (IAM) user, at this point. (See Chapter 13 in the *Ruby on Rails Tutorial* (https://www.railstutorial.org/book/user_microposts#sec-image_upload_in_production) for more information.)

4. Finally, make sure you're running an up-to-date version of Git (Listing A.1).

Because using two spaces for indentation is a near-universal convention in Ruby, I also recommend changing the editor to use two spaces instead of the default four. As shown in Figure A.7, you can do this by clicking the gear icon in the upper right and then changing the Soft Tabs setting to 2. (Note that this takes effect immediately; you don't need to click a Save button.)

---

3. Due to the constantly evolving nature of sites like AWS, details may vary; use your technical sophistication (Box A.1) to resolve any discrepancies.

4. https://aws.amazon.com/

5. https://www.railstutorial.org/cloud9-signup

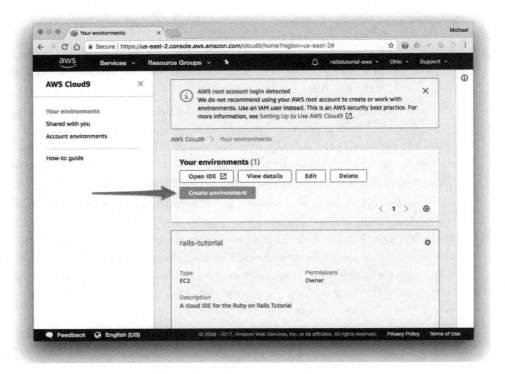

Figure A.3: The administrative page for Cloud9.

Listing A.1: Upgrading Git (if necessary).

```
$ git --version
# If the version number isn't greater than 2.28.0, run the following command:
$ source <(curl -sL https://cdn.learnenough.com/upgrade_git)
```

It's also important to use a compatible version of Node.js (https://nodejs.org/en/):

```
$ nvm install 16.13.0
$ node -v
v16.13.0
```

Finally, as of this writing the Learn Enough tutorials standardize on Ruby 2.7.4, which you can install on the cloud IDE as follows:

```
$ rvm install 2.7.4
```

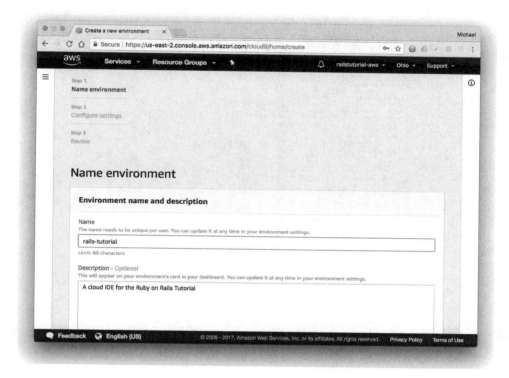

Figure A.4: Creating a new work environment at AWS Cloud9.

(This uses Ruby Version Manager (https://rvm.io/), which comes preinstalled on the cloud IDE.) Once that command is finished, you can verify the Ruby version as follows:

```
$ ruby -v
ruby 2.7.4p191 (2021-07-07 revision a21a3b7d23) [x86_64-linux]
```

(Exact version numbers may differ.)

At this point, you're done! Although Internet access is required to use Cloud9, there is no alternative that combines so much power with such an easy setup.

Figure A.5: Selecting **Ubuntu Server**.

A.3   Native OS Setup

As mentioned in Section A.1, setting up your native operating system as a development environment can be challenging, but it is an important step to take once you've reached a certain level of technical sophistication. The cloud IDE option is a great place to start, but eventually you have to grab the bull by the horns (Figure A.8)[6] and bend your native system to your will.

Section A.3.1 covers the conversion of macOS to a fully equipped development environment, while Section A.3.2 does the same for Linux. We cover Microsoft Windows options in Section A.3.3, but (as mentioned briefly in Section A.1) this section currently defers to the cloud IDE option in Section A.2.

---

6. Image courtesy of Rafael Ben-Ari/Alamy Stock Photo.

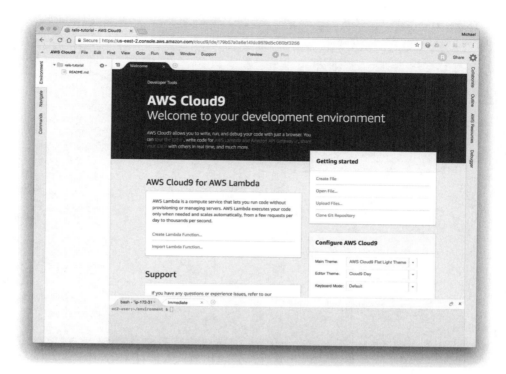

Figure A.6: The default cloud IDE.

A.3.1   macOS

The native Macintosh operating system, originally called Mac OS X and now known simply as *macOS,* has a polished graphical user interface (GUI) while being built on a solid Unix foundation. As a result, macOS is ideally suited for use as a programmer's development environment.

The steps in this section constitute more than just a minimal system; you can actually get away with doing a lot less, but your three authors all use macOS themselves, and we feel that it's important not to shortchange you with a half-baked setup.

Terminal and Editor

Although macOS comes with a native terminal program, we recommend installing iTerm (https://iterm2.com/downloads.html), which includes various enhancements that make it better than the default for developers and other technical users.

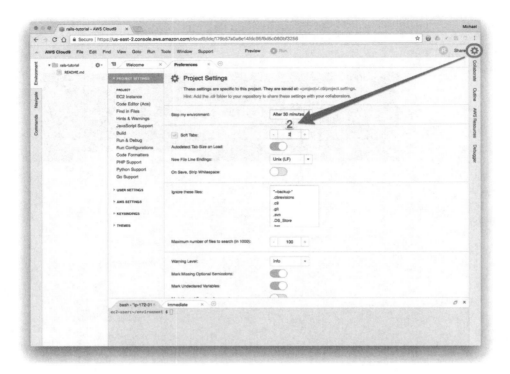

Figure A.7: Setting Cloud9 to use two spaces for indentation.

We also recommend installing a programmer's text editor. There are lots of excellent choices, but the Atom editor (https://atom.io/) (covered in *Learn Enough Text Editor to Be Dangerous*) is a good place to start if you don't already have a favorite.

By the way, for the Learn Enough tutorials and the Rails Tutorial, it is generally recommended that you use the Bourne-again shell (Bash) rather than the default Z shell (although it often doesn't matter). To switch your shell to Bash, run **chsh -s /bin/bash** at the command line, enter your password, and restart your terminal program. Any resulting alert messages are safe to ignore. See the Learn Enough blog post "Using Z Shell on Macs with the Learn Enough Tutorials" (https://news.learnenough.com/macos–bash–zshell) for more information.

Figure A.8: Sometimes you have to grab the bull by the horns.

Xcode Command-Line Tools

Although based on Unix,[7] macOS doesn't ship with all the software necessary for a proper development environment. In order to fill this gap, macOS users should install *Xcode*, a large suite of development tools and code libraries created by Apple.

Xcode used to require a 4+ GB download of installation source files, but thankfully Apple has recently made Xcode incredibly quick and easy to install with a simple command-line command, as shown in Listing A.2.

Listing A.2: Installing Xcode command-line tools.

```
$ xcode-select --install
```

---

7. Specifically, the NeXT system developed by the company Steve Jobs founded in 1985 after being ousted from Apple. The NeXT OS became the foundation for Mac OS X (later, macOS) after Apple acquired NeXT in 1997, which also led to Jobs' triumphant return as Apple CEO.

Homebrew

The next step is technically optional, but in our view it is necessary for a truly professional-grade macOS dev environment: namely, installing the outstanding *Homebrew* package manager.

You can think of a package manager as an App Store that runs at the command line and is filled with free open-source software. Nowadays most Linux distributions come with a native package manager (Section A.3.2), but by default macOS doesn't have one. Homebrew is one of many managers that is available in the open-source community, but over time it has become the most popular option among serious macOS developers.

As we'll see in Section A.3.1, we'll be using a program called *rbenv* to install Ruby, which in turn will be installed via Homebrew in Section A.3.1. But Homebrew itself requires Ruby, so it seems like we might have run into a circular dependency. Happily, macOS ships with a *system Ruby* that we can use to bootstrap the installation. We'll use this default Ruby to install Homebrew, and then we'll install rbenv and additional Ruby versions as outlined above.

The Homebrew installation program is a Bash script (https://www .learnenough.com/text-editor-tutorial/advanced_text_editing#sec-writing_an_exe -cutable_script), and can be accessed using the **curl** program covered in Section 3.1. We can execute the Homebrew installation script using the system **bash** executable, which is located in the **/bin** directory. The full command appears as in Listing A.3.

Listing A.3: Installing the Homebrew package manager.

```
$ /bin/bash -c "$(curl -fsSL https://www.learnenough.com/homebrew.sh)"
```

Note that Listing A.3 uses a learnenough.com forwarding URL, which points to the current Homebrew installation script. This way, if Homebrew changes the URL used to host the script, we can simply update the forwarding address, and this tutorial will continue to work as written. (You can also just copy and paste the full command from the Homebrew home page (https://brew.sh/) if you like.)

Homebrew installs a **brew** command-line command used for installing, updating, and removing packages. After the installation of Homebrew finishes, it's a good idea to run **brew doctor**, which ensures that all of the directories and permissions needed by Homebrew to manage local files are correctly set up:

```
$ brew doctor
```

If you have any problems at this point, you'll need to refer to the Homebrew troubleshooting wiki (https://docs.brew.sh/Troubleshooting), but you really shouldn't unless you've been making changes to random system folders and permissions.

Ruby Environment (rbenv)

As we've just seen, macOS comes with Ruby preinstalled, but we don't have any control over the exact version, and macOS doesn't natively allow us to use multiple versions of Ruby in parallel. To give us more flexibility with our development environment, we'll install *rbenv*, which is a utility that manages different Ruby versions and makes sure that Ruby software packages (called *gems*) get placed in the right spot for Ruby to find.

Using rbenv together with the associated *ruby-build* command also allows us to specify a different version of Ruby for different project repositories, which is a common task in software development.[8] For example, an older version of a program might need an older version of Ruby to run correctly. Using rbenv means we can support such an older program while still running a more up-to-date version of Ruby for our other projects.

Installing rbenv is easy with Homebrew. We can install both rbenv and ruby-build in one step using **brew install rbenv**, as shown in Listing A.4.

Listing A.4: Installing rbenv and ruby-build.

```
$ brew install rbenv    # automatically installs ruby-build as well
```

After the installation from Listing A.4 finishes, we need to get rbenv up and running using **rbenv init**, as shown in Listing A.5.

Note: If you're using Zsh instead, substitute **.zshrc** for **.bash_profile** in this section. See "Using Z Shell on Macs with the Learn Enough Tutorials" for more information.

---

8. E.g., the *virtualenv* utility accomplishes a similar task for Python projects.

Listing A.5: Initializing rbenv.

```
$ rbenv init
# Load rbenv automatically by appending
# the following to ~/.bash_profile (or ~/.zshrc if using Zsh):

eval "$(rbenv init -)"
```

If Listing A.5 gives you an error message like "No such file or directory", try exiting your shell program with Ctrl-D and restarting it, and then try the command again. (This sort of restart-and-retry technique is classic technical sophistication (Box A.1).)

As seen in Listing A.5, running **rbenv init** gives us a suggestion for how to avoid having to initialize rbenv by hand each time: We simply need to append the line

```
eval "$(rbenv init -)"
```

to our Bash profile file **.bash_profile** (Section 5.4).

If you prefer, you can use a text editor to add the **eval** line to your system's **.bash_profile** file, but the easiest way is to use **echo** and the append operator **>>** covered in Section 2.1, like this:

```
$ echo 'eval "$(rbenv init -)"' >> ~/.bash_profile    # or ~/.zshrc if using Zsh
```

Note that we've included the home directory **~** in the path so that it works no matter which directory we're currently in.

Finally, to activate the new profile file we need to **source** it (as seen in Listing 5.5):

```
$ source ~/.bash_profile    # or ~/.zshrc if using Zsh
```

New Ruby Version

Now that rbenv is set up, let's give it a non–system version of Ruby to manage. The installation process is handled entirely by rbenv, so all you have to do is tell it which version you'd like on your system by passing along the exact Ruby version number.

We'll use Ruby 2.7.4 in this tutorial, which as of this writing works with a wide variety of Ruby applications, but you can also use a current or previous version as listed at the Ruby website.

To install the desired version of Ruby using **rbenv**, simply execute the command shown in Listing A.6. If your system ever complains that the given Ruby version isn't available, you'll have to update your system to access the latest version (Box A.2).

Listing A.6: Installing a fresh copy of Ruby.

```
$ rbenv install 2.7.4
```

After running the command in Listing A.6, you should see rbenv start the download process and install any dependencies that are needed for that specific version of Ruby (which might take a while depending on bandwidth and CPU limitations).

Note: You may get an error message of the form

```
ruby-build: definition not found: 2.7.4

See all available versions with `rbenv install --list`.

If the version you need is missing, try upgrading ruby-build:

  brew update && brew upgrade ruby-build
```

If this happens, you should clone the freshest **ruby-build** list into the rbenv **plugins** directory:

```
$ mkdir ~/.rbenv/plugins
$ git clone https://github.com/rbenv/ruby-build.git ~/.rbenv/plugins/ruby-build
```

If **ruby-build** is already there, you can simply **pull** in the latest changes:

```
$ cd ~/.rbenv/plugins && git pull && cd -
```

Box A.2: Updating and Upgrading

Although we just installed fresh versions of all relevant software, after a while your local system will get out of sync with the latest versions of Homebrew and any installed packages. In order to update the system, every once in a while it's a good idea to update Homebrew itself and then upgrade the installed packages as follows:

```
$ brew update
$ brew upgrade
```

After the Ruby installation finishes, we need to tell the system that there's a new version of Ruby using the obscurely named **rehash** command:

```
$ rbenv rehash
```

For this guide, we are also going to set the Ruby version from Listing A.6 as the global default so that you won't have to worry about specifying the Ruby version when you start your project. The way to do this is with the **global** command:

```
$ rbenv global 2.7.4
```

At this point, it's probably a good idea to restart your shell program to make sure all the settings are properly updated.

For future work, you may want to use specific versions of Ruby on a per-project basis, which can be done by creating a file called **.ruby-version** in the project's root directory and including the version of Ruby to be used. (If it isn't already present on your system, you'll also have to install it using **rbenv install <version number>** as in Listing A.6.) See the rbenv documentation (https://github.com/rbenv/rbenv) for more information.

Finally, when installing Ruby software via *gems*, or self-contained packages of Ruby code, it's often convenient to skip the installation of the local documentation, which can take more time to install than the software itself, and in any case is more conveniently accessed online. Preventing documentation installation can be done on a case-by-case basis, but it's more convenient to make it a global default by creating a file in the home directory called **.gemrc** with instructions to skip the rarely used (and time-consuming to install) Ruby documentation files, as shown in Listing A.7.

Listing A.7: Configuring the **.gemrc** file to skip the installation of Ruby documentation.

```
$ echo "gem: --no-document" >> ~/.gemrc
```

With that configuration, any uses of **gem install <gem name>** to install Ruby gems will automatically be svelte, streamlined, and documentation-free.

Git

A recent version of the Git version control system should come automatically with the Xcode command-line tools installed in Section A.3.1, which you can verify using the **which** command:

```
$ which git
```

If the result of this is blank, it means Git isn't installed, and you can install it with Homebrew:

```
$ brew install git
```

Either way, you should check to make sure that the Git version number is at least **2.28.0**:

```
$ git --version
git version 2.31.1    # should be at least 2.28.0
```

If the version number isn't recent enough, you should run **brew update** and then run **brew upgrade git**.

A.3.2 Linux

Because of Linux's highly technical origins, Linux systems typically come well-equipped with developer tools. As a result, setting up a native Linux OS as a dev environment is especially simple.

Every major Linux distribution ships with a terminal program, a text editor, and Git. There are only four major steps we recommend in addition to the defaults:

1. Download and install Atom if you don't already have a favorite editor.
2. Follow the rbenv installation instructions (https://github.com/rbenv/rbenv# installation) from the rbenv website.
3. Install and configure Ruby as shown in Listing A.6 and Listing A.7.
4. Make sure that the Git version number is at least **2.28.0**.

The final step can be accomplished by running

```
$ git --version
```

If the output isn't at least **2.28.0**, go to "Getting Started – Installing Git" (https://git-scm.com/book/en/v2/Getting-Started-Installing-Git) and install the latest version for your system.

At this point, you should be good to go!

A.3.3 Windows

Finally, we have native instructions for Microsoft Windows—or, rather, instructions for using Unix on Windows. As discussed in Section A.2, one possibility is to use a cloud IDE, but we've had reports of especially good results with installing Linux directly in Windows.

That's right! Believe it or not, Windows now ships with a working Linux kernel, and you can install any of a number of Linux distributions by following Microsoft's own instructions. (To those of us who remember the Linux-hating, predatory Microsoft of the late '90s/early 2000s, the idea that Windows would someday ship with native Linux support is truly incredible—*dogs and cats, living together!*—and yet here we are.)

To get everything working, our recommendation is to follow the tutorial article "Ruby on Rails on Windows is not just possible, it's fabulous using WSL2 and VS Code" (https://www.hanselman.com/blog/ruby-on-rails-on-windows-is-not-just-possible-its-fabulous-using-wsl2-and-vs-code) by Scott Hanselman (https://www.hanselman.com/). Although especially useful for Ruby on Rails web development, Hanselman's tutorial is of quite general applicability, and following it should result in a fantastic general development experience on Windows.

A.4 Conclusion

If you've made it this far—and especially if you completed a native OS setup in Section A.3—you've now learned enough dev environment to be *dangerous*. You're ready to move on to complete challenging tutorials like *Learn Enough CSS & Layout to Be Dangerous* (https://www.learnenough.com/css-and-layout), *Learn Enough Ruby to Be Dangerous*, and the *Ruby on Rails Tutorial*. Good luck!

Index

Photo by Marvent/Shutterstock

VIDEO TRAINING FOR THE **IT PROFESSIONAL**

LEARN QUICKLY
Learn a new technology in just hours. Video training can teach more in less time, and material is generally easier to absorb and remember.

WATCH AND LEARN
Instructors demonstrate concepts so you see technology in action.

TEST YOURSELF
Our Complete Video Courses offer self-assessment quizzes throughout.

CONVENIENT
Most videos are streaming with an option to download lessons for offline viewing.

Learn more, browse our store, and watch free, sample lessons at
informit.com/video

Save 50%* off the list price of video courses with discount code **VIDBOB**

inform IT ®
the trusted technology learning source